Value Distribution Theory

THE UNIVERSITY SERIES IN HIGHER MATHEMATICS

HALMOS, PAUL R.—Measure Theory
JACOBSON, NATHAN—Lectures in Abstract Algebra
 Vol. I—Basic Concepts *Vol. II*—Linear Algebra
 Vol. III—Theory of Fields and Galois Theory
KLEENE, S. C.—Introduction to Metamathematics
LOOMIS, LYNN H.—An Introduction to Abstract Harmonic Analysis
LOÉVE, MICHEL—Probability Theory, 3rd Edition
KELLEY, JOHN L.—General Topology
ZARISKI, OSCAR, and SAMUEL, PIERRE—Commutative Algebra, Vols. I and II
GILLMAN, LEONARD, and JERISON, MEYER—Rings of Continuous Functions
RICKART, CHARLES E.—General Theory of Banach Algebras
J. L. KELLEY, ISAAC NAMIOKA, and CO-AUTHORS—Linear Topological Spaces
SPITZER, FRANK—Principles of Random Walk
NACHBIN, LEOPOLDO—The Haar Integral
KEMENY, JOHN G., SNELL, J. LAURIE, and KNAPP, ANTHONY W.—Denumerable Markov Chains
SARIO, LEO, and NOSHIRO, KIYOSHI—Value Distribution Theory

A series of advanced text and reference books in pure and applied mathematics. Additional titles will be listed and announced as published.

Value Distribution Theory

LEO SARIO

Professor of Mathematics
University of California
Los Angeles, California

AND

KIYOSHI NOSHIRO

Professor of Mathematics
Nagoya University
Nagoya, Japan

in collaboration with

TADASHI KURODA　　KIKUJI MATSUMOTO　　MITSURU NAKAI

D. VAN NOSTRAND COMPANY, INC.

PRINCETON, NEW JERSEY

TORONTO　　NEW YORK　　LONDON

D. VAN NOSTRAND COMPANY, INC.
120 Alexander St., Princeton, New Jersey (*Principal office*)
24 West 40 Street, New York 18, New York

D. VAN NOSTRAND COMPANY, LTD.
358, Kensington High Street, London, W.14, England

D. VAN NOSTRAND COMPANY (Canada), LTD.
25 Hollinger Road, Toronto 16, Canada

SOFTCOVER REPRINT OF THE HARDCOVER 1ST EDITION 1966

Published simultaneously in Canada by
D. VAN NOSTRAND COMPANY (Canada), LTD.

ISBN 978-1-4615-8128-4 ISBN 978-1-4615-8126-0 (eBook)
DOI 10.1007/978-1-4615-8126-0

DEDICATED TO

The University of California
which gave us the opportunity to discover

CONTENTS

Chapter II MAPPINGS INTO OPEN RIEMANN SURFACES

Chapter III FUNCTIONS OF BOUNDED CHARACTERISTIC

Chapter IV FUNCTIONS ON PARABOLIC RIEMANN SURFACES

ACKNOWLEDGMENTS

We are deeply grateful to the U. S. Army Research Office–Durham, in general, and to Drs. John W. Dawson and A. S. Galbraith in particular, for several U.C.L.A. grants during the five years 1961–1966 which the writing of the book has taken. Were it not for their patience with our everchanging plans, this work may never have been completed.

Our sincere thanks are due to Professor S. S. Chern for the inclusion of our book in this distinguished series and for his continued stimulation.

We are indebted to many colleagues who read the manuscript, in particular our collaborator M. Nakai, who made substantial contributions to several parts of the theory and scrutinized the entire manuscript; our collaborator K. Matsumoto, who contributed his conclusive results on Picard sets; our collaborator T. Kuroda, who with Matsumoto and Nakai covered an early version of the manuscript in a seminar; our esteemed friend L. Ahlfors, whose council we had the advantage of obtaining on several occasions; K. V. R. Rao, who helped us with the second half of Chapter III; B. Rodin, who made valuable suggestions; M. Glasner, who compiled the Indices and assisted us with the numerous tasks of preparing the manuscript for printing; P. Emig and S. Councilman, who with Glasner compiled the Bibliography.

We were fortunate to have the typing of the several versions of the manuscript in the expert hands of Mrs. Elaine Barth and her efficient staff.

PREFACE

The purpose of this research monograph is to build up a modern value distribution theory for complex analytic mappings between abstract Riemann surfaces. All results presented herein are new in that, apart from the classical background material in the last chapter, there is no overlapping with any existing monograph on meromorphic functions.

Broadly speaking the division of the book is as follows: The Introduction and Chapters I to III deal mainly with the theory of mappings of arbitrary Riemann surfaces as developed by the first named author; Chapter IV, due to Nakai, is devoted to meromorphic functions on parabolic surfaces; Chapter V contains Matsumoto's results on Picard sets; Chapter VI, predominantly due to the second named author, presents the so-called nonintegrated forms of the main theorems and includes some joint work by both authors. For a complete list of writers whose results have been discussed we refer to the Author Index.

The value distribution theory had its inception with Picard's celebrated theorem, one of the most beautiful results in classical analysis. It was the starting point of the pioneering work of the French school: Borel, Hadamard, Valiron, and Julia. In 1924 Collingwood and Littlewood made a fundamental discovery: there can be infinitely many defectively covered points, and the defect sum (for entire functions) cannot exceed 1. This Collingwood-Littlewood defect relation is still the cornerstone of value distribution theory, and the primary object of the fundamental theorems. The relation was generalized by Nevanlinna, who also introduced the present terminology: "counting function" for the function initiated by Valiron to describe the coverage of a point, "characteristic function" for the Valiron function for fully covered points, and "proximity function" for the deviation of the former from the latter. The most effective methods currently in use, both for the integrated and nonintegrated forms, were devised by Ahlfors. To Nevanlinna is due the development of value distribution theory into a beautiful unity, a masterpiece in the art of mathematics.

Despite its elegance the classical theory suffers from the following restriction: a meromorphic function is a locally defined concept, and its potentialities are curtailed by confining it to a globally chosen special carrier such as the plane or the disk. Full richness of the theory can be expected only on the corresponding locally defined carrier, the most general one on which the concept of analytic function makes sense.

Building up a general theory of complex analytic mappings between Riemann surfaces thus appears to be of compelling importance.

In the following sense we have reached this goal: we have established the integrated forms of the main theorems for analytic mappings into both closed and open Riemann surfaces, and the nonintegrated forms for mappings into closed surfaces. Moreover, we have obtained a bound for the number of Picard points for mappings into closed surfaces, and have shown that the capacity of the set of exceptional points vanishes for mappings into arbitrary surfaces, closed or open. In contrast, we have only fragmentary results to report on the existence of mappings between given surfaces. In this direction the road is open for further research.

Nowhere in the book have we made any attempt at completeness. We have been led mainly by our own interests and a desire for natural unity.

The Introduction is intended to orient the reader with the techniques we have used, in particular, the new proximity function $s(\zeta, a)$ and the method of areal proximity as compared with the classical curvilinear proximity. These tools permit us to obtain in Chapter I the main theorems for analytic mappings of *arbitrary* Riemann surfaces. In contrast with the classical theory our forms of the main theorems are valid for all subregions, with no exceptional ones omitted. As an extension of an elegant result by Chern we obtain the bound $\eta - e_S$ for the number of Picard points.

The class of R_p-surfaces is then introduced. It is characterized by the existence of capacity functions p with compact level lines. In Chapter I these surfaces provide us with the strictness of our bound for the number of Picard values and with a short proof of Nevanlinna's classical defect relation.

Chapter II opens with the important theory of principal functions, indispensable in Chapters I to IV, and VI. In Chapter II these functions are used to establish the uniform boundedness from below of $s(\zeta, a)$ in both variables for an arbitrary Riemann surface. This result leads to the main theorems for given nondegenerate mappings into arbitrary Riemann surfaces and to the affinity relation for such mappings into surfaces of finite Euler characteristic. It also gives, as shown by Nakai, the joint continuity of $s(\zeta, a)$ and the vanishing of the capacity of the set of exceptional points under all mappings of arbitrary Riemann surfaces. Rodin's and Ozawa's results on the existence of analytic mappings are included.

Chapter III starts with a decomposition theorem for meromorphic functions of bounded characteristic on arbitrary Riemann surfaces. The class O_{MB} of surfaces without such functions is then studied, and theorems of Heins, Parreau, and Rao on the decomposition into quasi-bounded and singular parts are established.

Chapter IV contains Nakai's penetrating proof, using the Čech compactification, of the existence of the Evans-Selberg potential on arbitrary parabolic surfaces. This solution of the long open problem places parabolic surfaces in the class of R_p-surfaces. For functions on such surfaces it is shown that the set of exceptional values has vanishing capacity, an extension of the af Hällström-Kametani-Nevanlinna theorem.

That this theorem is sharp is the striking result of Matsumoto given in Chapter V: for every compact set K of vanishing capacity there exists a meromorphic function with a set E of essential singularities of vanishing capacity and with exactly K as the Picard set at every point of E. A necessary condition is then obtained for every meromorphic function in the complement of a given E to have a finite Picard set at each point of E. The theorem is a sharpening of a recent interesting result of Carleson.

The longest chapter is VI. It starts with Ahlfors' elegant theory of covering surfaces, which is used to derive the nonintegrated forms of the main theorems on analytic mappings of arbitrary Riemann surfaces into closed surfaces. Corresponding theorems are also proved in a form localized to a transcendental singularity of the inverse function.

At the end of Chapter VI we once more return to R_p-surfaces and show, generalizing an idea of Dinghas, that integrated forms of the main theorems can be derived from the nonintegrated forms. The general case of R_p-surfaces is then compared with the important special case of algebroids.

To avoid interrupting the train of thought in value distribution theory proper, we have compiled in Appendix I some basic properties of Riemann surfaces that are referred to in Chapters I to VI.

In Appendix II we first give an explicit construction of complete minimal surfaces of arbitrary connectivity and genus, smoothly immersed in E^3. Although this construction, due to Klotz and the first named author, is somewhat isolated from the rest of the book, we believe that the value distribution theory of Gaussian mappings of these surfaces illuminates the general theory since the fundamental quantities assume concrete meanings. Taking the proximity function into account, neglected thus far, should be fruitful in further research on Gaussian mappings.

Beyond the above broad description of the book we have given a more detailed orientation on its plan and interconnections in the introductions to the chapters, sections, and appendices. The reader will do well to read them before starting a systematic study.

The expert will note that several results not previously published are scattered through the book, and earlier ones or their proofs are improved. For example, Theorems I.2E and II.9A now appear without remainders and simplify the entire theory considerably. However, no changes have been made for the sake of changes. Where the authors had no improvements to report, their original presentations have been followed rather closely.

The reader is not expected to have any previous knowledge of value distribution theory. For general prerequisites a standard Ph.D. curriculum in complex analysis, real analysis, functional analysis, topology, differential geometry, and algebra should be sufficient. Outside of this we have in a few instances quoted well-known theorems if their proofs are easily obtainable from an established source; exact reference is then made.

Bibliographical references concerning main results are placed in the introductions to chapters and sections, and again at the theorems. Sometimes "Remarks" are used for this purpose.

Cross-references are self-explanatory: e.g., I for Chapters, I. §1 for sections, I.1 for numbers, I.1A for subnumbers, I.(1) for formulas, and Theorem I.2E for statements. For the convenience of the reader we have occasionally repeated some definitions and related preliminaries.

For comprehensive studies of the main tools used in this book we refer to the forthcoming monographs Rodin-Sario, "Principal functions" (to appear in this series); Oikawa-Sario, "Capacity functions"; and Nakai-Sario, "Classification theory".

In accordance with the plan of our book we have only lightly touched on the classical theory of meromorphic functions as presented in Nevanlinna's French monograph on the Picard-Borel theorem and later in his well-known German treatise. In this classical direction far-reaching further results have been obtained by Hayman and Edrei-Fuchs in their monographs on meromorphic functions and by Matsumoto in Chapter V of the present book.

At the end of our book a rather comprehensive bibliography on value distribution theory, classical and modern, is published for the first time. (For reasons of space, literature on entire functions and other more restricted topics was generally not included.) It is our hope that the bibliography will be useful to workers in the field. It also reveals the profound influence Picard's theorem and the Collingwood-Littlewood defect relation have had on the evolution of complex analysis.

Los Angeles, California
April 1, 1966

LEO SARIO
KIYOSHI NOSHIRO

INTRODUCTION

The purpose of this Introduction is to give, without proofs, a general framework into which our method will be built in Chapters I and II. For a reader with some previous knowledge of value distribution theory and the theory of Riemann surfaces it also offers a comparative survey of the classical and the new approach.

1. Historical. During the nine decades that have elapsed since the publication of Picard's theorem, evolution has taken place toward greater generality: the Picard-Borel-Nevanlinna theory was first extended from the plane to more general plane regions by af Hällström [2] and Tsuji [7], then to various Riemann surfaces by Ahlfors [11], [13], Heins [1], Kunugui [3], Kuramochi [1], L. Myrberg [1], Noshiro [5], Ohtsuka [3], Parreau [2], Tamura [1], Tsuji [15], Tumura [4], and others.

The most general result was obtained in 1960 by Chern [1], who considered as domain R a closed Riemann surface less a finite number of points, and as range S a closed Riemann surface. He showed that under a nondegenerate complex analytic mapping $\zeta = f(z)$ of R into S, z and ζ being the local complex parameters, the number P of Picard points, and more generally the defect sum, cannot exceed the negative of the Euler characteristic of S:

$$P \leq -e_S. \tag{1}$$

This beautiful result of Chern's paves the way for the following question: Can generality be pushed further by allowing both R and S to be arbitrary? A priori this did not seem likely. In fact, Heins [1] had exhibited an interesting Riemann surface of infinite genus, which carried meromorphic functions with infinitely many Picard values. A look in a somewhat different direction reveals, however, rather interesting new aspects. To this end let us consider current methods and see if we can introduce simplifications which permit greater generality.

2. New metric. The first tool we need is a function to describe the proximity of a generic point $\zeta \in S$ to a given point $a \in S$. The standard method is the following: one first forms a conformal metric with area element $d\omega = \lambda^2 \, dS$, where dS is the Euclidean area element in the parametric disk, and λ is covariant and strictly positive. Throughout our presentation let $t(\zeta, a, b)$ be a harmonic function of the variable ζ on S with a positive logarithmic pole at a and a negative logarithmic pole at b. One integrates t with respect to $d\omega(b)$ over S. The resulting function

$$q(\zeta, a) = \int_S t(\zeta, a, b)\, d\omega(b) \tag{2}$$

is bounded from below and has a positive logarithmic singularity at a. It thus qualifies to describe the proximity of ζ to a. Moreover, Δq is simply the "density" of the metric:

$$\Delta q = \lambda^2. \tag{3}$$

This makes it possible to use effectively the standard relations between line and area integrals.

There are, however, two drawbacks to this approach. First, if S is open, it seems difficult, if not impossible, to establish the convergence of integral (2). Second, even when S is closed, a rather lengthy reasoning in partial differential equations is needed to show that Δq actually is λ^2. If S is open, there seems to be no way of putting the reasoning through.

To overcome this difficulty we suggest the following reversal of the process: start with a function

$$t_0(\zeta) = t(\zeta, \zeta_0, \zeta_1) \tag{4}$$

with given $\zeta_0, \zeta_1 \in S$. The singularities together with a normalization of the additive constant uniquely determine t_0 if S is closed. If S is bordered and compact, then we add the condition that the function be constant on the border. If S is open, we take for t_0 the directed limit of the functions thus constructed on bordered subregions as the subregions exhaust S. The limiting function is a special case of the so-called principal function, and its existence is assured by the related linear operator method (Sario [1]). The function

$$s_0(\zeta) = \log (1+e^{t_0(\zeta)}) \tag{5}$$

is bounded from below but continues to have a positive logarithmic pole at ζ_0. For any other point a take $t(\zeta, a, \zeta_0)$ and add it to $s_0(\zeta)$. The singularities at ζ_0 cancel and the function

$$s(\zeta, a) = s_0(\zeta)+t(\zeta, a, \zeta_0) \tag{6}$$

is bounded from below and has a positive logarithmic singularity at a. We choose this function to describe the proximity of ζ to a: closer proximity gives greater values. The function exists on every Riemann surface S, open or closed, of finite or infinite genus.

Having formed s we introduce a conformal metric with area element $\lambda^2\, dS$ by choosing the density $\lambda^2 = \Delta s = \Delta s_0$. It is independent of a. Thus the problem of convergence of (2) and the proof of (3) are eliminated, and the metric is obtained on an arbitrary S.

In this metric λ has zeros which are those of grad t_0. But these zeros turn out to be helpful and, in fact, constitute a rather essential aspect of the theory.

In passing we remark that the Gaussian curvature of our metric is constantly 1 and its total area is $\int_S d\omega = 4\pi$. As a by-product we thus have a conformal metric (which has zeros of λ) of constant curvature and finite total area on an arbitrary Riemann surface.

3. The fundamental A-, B-, and C-functions. We can now at once write down the first main theorem; it was earlier considered from different viewpoints by Heins [3], Kuramochi [1], L. Myrberg [1], and Parreau [2]. Here we give it in a form that directly serves the first purpose of this book: Picard's theorem on an arbitrary R.

Remove from R a parametric disk R_0 with boundary β_0, and consider an adjacent regular region $\Omega \subset R$ with boundary $\beta_0 \cup \beta_\Omega$. On $\bar{\Omega}$ form the harmonic function u with $u=0$ on β_0, $u=k$, a constant, on β_Ω, such that the flux $\int_{\beta_0} du^* = 1$. For $h \in [0, k]$ consider the level line $\beta_h = u^{-1}(h)$ and the region $\Omega_h = u^{-1}((0, h))$ between β_0 and β_h. Given a point $a \in S$ let $\{z_j\}$ be its inverse images under f and denote their number in $R_h = \bar{R}_0 \cup \Omega_h$ by $\nu(h, a)$.

For the a-points we introduce the *A-function*

$$A(h, a) = 4\pi \int_0^h \nu(h, a)\, dh. \tag{7}$$

It reflects the frequency of the a-points of f on R. In particular, it vanishes for a Picard point a, i.e., a point which is not covered by f. For the β-curves we take the *B-function*

$$B(h, a) = \int_{\beta_h - \beta_0} s(f(z), a)\, du^*. \tag{8}$$

Its geometric meaning is clear: it is the mean proximity to a of the image of $\beta_h - \beta_0$ under f. Finally, the growth of the image area is characterized by the *C-function*

$$C(h) = \int_0^h \int_{R_h} d\omega(f(z))\, dh, \tag{9}$$

with $C'(h)$ the area $\int_{R_h} d\omega(f(z))$ of the (multisheeted) Riemannian image under f of R_h over S.

A simple application of Stokes' formula to Ω_h less small disks about the z_j that shrink to their centers gives the following

Theorem. *For every regular region $\Omega \subset R$ under an analytic mapping of an arbitrary Riemann surface R into another arbitrary Riemann surface S,*

$$A(k, a)+B(k, a) = C(k). \tag{10}$$

Thus the elegant classical balance prevails: the $(A+B)$-affinity, so to speak, of f is the same for all points $a \in S$. In particular, for a Picard point a, $A \equiv 0$ and we have a strong proximity of $f(\beta_h - \beta_0)$ to a.

4. Method of areal proximity. We now come to the main question: How many Picard points $a_1, \cdots, a_q$ can there exist? The answer is given by the second main theorem which we shall here give for mappings of an arbitrary R into a closed S.

It is well known that in the classical second main theorem the remainder cannot be estimated for all values of the variable r. It is the integral of the integral of the remainder that can be given a dominating function. The remainder itself can behave arbitrarily wildly in certain intervals whose length can be estimated but which must be omitted in stating the second main theorem. When one then takes the defect relation, these exceptional intervals and the related changing of the coordinate system with varying Ω prevent the use of directed limits. But ordinary limits cannot be employed on an *arbitrary* Riemann surface R: there is no one parameter that would give an exhaustion of R. Thus the classical theory does not carry over to the general case.

This difficulty can be overcome by the following simple device. We replace the proximity function B by the integral of its integral. Geometrically the first integration means that we replace the mean proximity of the image *curve* $f(\beta_h)$ by what is just as natural if not more so, the mean proximity of the image *region* $f(\Omega_h)$, and then we take the integral of this. Analytically this means that, in some sense, we bring all quantities involved to the same level of integration. Then the remainder term in the second main theorem has an estimate for every subregion Ω, directed limits can be employed, and the theory established on an arbitrary R.

The actual derivation of the second main theorem consists of little more than another application of Stokes' formula. The proof is further facilitated by the presence of the zeros of λ to which we referred earlier. Their number is the Euler characteristic of the punctured (at ζ_0 and ζ_1) S and we obtain e_S without using the Gauss-Bonnet formula. Geometrically this makes it unnecessary to set up the tangent bundle, and we can dispense with borrowing from differential geometry.

When the computations are carried out we obtain the following result:

Theorem. *For every $\Omega \subset R$ under an analytic mapping of an arbitrary R into a closed S,*

$$(q+e_S)C_2(k) < \sum_{i=1}^{q} A_2(k, a_i) - A_2(k, f') + E_2(k) + O(k^3 + k^2 \log C(k)), \tag{11}$$

where $C_2(k)$ is the integral of the integral from 0 to k of $C(h)$, $A_2(k, f')$ counts the multiple points of $f(\Omega)$, and $E_2(k)$ is the (4π-fold) thrice integrated Euler characteristic $e(h)$ of Ω_h.

The remainder term O is negligible for the nondegenerate class of functions which was given the following simple characterization by Rao: there must exist a constant $0 < \alpha < 1$ such that for $R_k = \bar{R}_0 \cup \Omega$

$$\lim_{R_k \to R} \frac{\log C(k)}{C(\alpha k)} = 0. \tag{12}$$

For these functions we can now introduce the defect

$$\alpha(a) = 1 - \limsup_{R_k \to R} \frac{A_2(k, a)}{C_2(k)}. \tag{13}$$

For a Picard point a this defect is obviously $=1$. We also introduce the ramification index

$$\beta(a) = \liminf_{R_k \to R} \frac{A_2(k, f'_a)}{C_2(k)}, \tag{14}$$

where $A_2(k, f'_a)$ counts the orders of the branch points above a, and what could be called the Euler index

$$\eta = \liminf_{R_k \to R} \frac{E_2(k)}{C_2(k)}. \tag{15}$$

We obtain at once:

Defect and ramification relation. *For nondegenerate analytic mappings of an arbitrary Riemann surface R into a closed Riemann surface S,*

$$\sum \alpha(a) + \sum \beta(a) \leq \eta - e_S. \tag{16}$$

We can now throw some light onto the Heins phenomenon. If C_2 grows less rapidly than E_2, then $\eta = \infty$ and there can be infinitely many Picard points. However, even in the elementary case of the disk R, e.g., the identity mapping omits infinitely many points, in fact the entire complement of R. The problem of Picard values becomes interesting only if a growth condition is imposed upon the characteristic function. For the disk such a condition is well known. For an arbitrary Riemann surface we now have, in addition to the nondegeneracy condition $\log C(k)/C(\alpha k) \to 0$, the essential condition reflecting the topology of R: *the characteristic function must grow at least as rapidly as the Euler characteristic.* For these mappings we have what we set out to find, a Picard theorem on an arbitrary R: the number of Picard points cannot exceed the excess of η over e_S,

$$P \leq \eta - e_S. \tag{17}$$

More accurately, $P \leq \eta - e_S - \sum \beta(a)$.

In the case of the sphere S, i.e., for meromorphic functions on arbitrary Riemann surfaces, the bound $2+\eta$ was shown to be sharp by an interesting example constructed by Rodin [2]. In the classical case of meromorphic functions in the plane we have an elementary proof of the defect relation, and a second main theorem without exceptional intervals.

5. Summary. The above is the approach our book starts with. To summarize, the advantages of the method are as follows:

(a) The most cumbersome part of the reasoning, that involving exceptional intervals, is eliminated.

(b) The resulting "second main theorem" is valid for all subregions, with no "exceptional" ones omitted.

(c) The degeneracy of mappings is characterized in a simple and uniform manner for all surfaces.

(d) The results are obtained simultaneously in all cases, without the necessity of distinguishing between sup $k=\infty$ and sup $k<\infty$. In particular, meromorphic functions in the plane and in the disk are treated at once.

(e) Even the classical theorems in these special cases are more effectively derived as consequences of the general theorems.

(f) These results are valid for meromorphic functions on *arbitrary* Riemann surfaces.

(g) The results are not restricted to meromorphic functions but apply to mappings into arbitrary closed Riemann surfaces as well.

(h) The method can be largely extended to the most general case of complex dimension 1: given mappings of arbitrary Riemann surfaces into arbitrary Riemann surfaces.

Remark. The above Introduction is, in essence, an invited lecture "Complex analytic mappings" delivered before a meeting of the American Mathematical Society (Sario [12]).

CHAPTER I

MAPPINGS INTO CLOSED RIEMANN SURFACES

A Riemann surface R is a connected Hausdorff space with a conformal structure (cf. e.g., Ahlfors-Sario [1, p. 114]). We shall use the same symbol z for a generic point and its parametric image. Let S be another Riemann surface, and ζ its parameter. A mapping $\zeta = f(z)$ of R into S is by definition analytic if it is so in terms of the parameters. We are interested in the distribution of values of f. Typically we ask: How many points of S can f omit?

Using the time-honored terminology we shall call a Riemann surface closed or open according as it is compact or not. We shall proceed to full generality of the main theorems in two steps: in Chapter I we consider analytic mappings into closed surfaces; in Chapter II, those into arbitrary surfaces. This arrangement will permit us to clearly bring forth in Chapter I the essentials of our method, and the somewhat delicate reasoning on the proximity function in the general case can be postponed to the beginning of Chapter II. Proofs of peripheral intuitively clear steps in Chapter I (e.g., in 1B, 1C) can also be relegated to the corresponding passages of Chapter II (7D, 8C, resp.). The slight overlapping of the chapters will only facilitate the access to the main results in Chapter II.

The general theory of the present chapter will be developed in §1. The above question on omitted points is answered in Theorem 7A, which is a consequence of our main result, Theorem 5C for arbitrary domain surfaces. In §2 the general theory is applied to the case of the sphere S, i.e., meromorphic functions on arbitrary Riemann surfaces R. For this case we give in §3 another proof along more classical lines, partly for completeness, partly for comparison. The proof involves exceptional intervals and we are restricted to a special class R_s of domain surfaces.

In the Introduction we listed earlier literature, essential for later work. For the method developed in the present chapter for analytic mappings of arbitrary Riemann surfaces we refer to Sario [4], [5], [6], [9], and [12]. The presentation here will be self-contained.

§1. MAPPINGS OF ARBITRARY RIEMANN SURFACES

In this section we shall first introduce the proximity function $s(\zeta, a)$, the basic tool in our approach. We then define the fundamental A-, B-, and

C-functions and derive the main theorems 2E and 5C governing them. As consequences we obtain the defect and ramification relations for analytic mappings of arbitrary open Riemann surfaces into arbitrary closed Riemann surfaces.

Some terminology is taken from the theory of Riemann surfaces. A subregion of an open Riemann surface is called *regular* if its closure is compact and its boundary consists of a finite number of analytic Jordan curves. An axiomatic treatment of this and related concepts is to be found, e.g., in Ahlfors-Sario [1].

1. The proximity function $s(\zeta, a)$

1A. Let f be an analytic mapping of an arbitrary Riemann surface R into a closed Riemann surface S. Value distribution theory deals with the affinity (to be specified) of f with respect to given points $a_1, \cdots, a_q$ on S.

First we shall construct a proximity function on S, i.e., a function to measure the nearness of a generic point ζ to a given point a. Choose ζ_0, ζ_1 on S, different from the a_i, $i=1, \cdots, q$, and take arbitrary but then fixed disjoint parametric disks D_0, D_1 about ζ_0, ζ_1. Let t_0 be a harmonic function in $S-\zeta_0-\zeta_1$ with singularities $-2 \log |\zeta-\zeta_0|$ and $2 \log |\zeta-\zeta_1|$ in D_0 and D_1, respectively. The existence of such a function on a closed S is classical. (For the general case of a closed or open S the construction of t_0 is carried out in II.6A.) We normalize the additive constant by the condition $t_0(\zeta)+2 \log |\zeta-\zeta_0| \to 0$ as $\zeta \to \zeta_0$. The function

$$s_0(\zeta) = \log (1+e^{t_0(\zeta)}) \tag{1}$$

continues to have a positive logarithmic pole at ζ_0 and is nonnegative on S. It is our proximity function for ζ_0: greater proximity gives greater values.

1B. For any other point $a \neq \zeta_0$ we could form the proximity function in the same manner. But we wish it to have the same Laplacian as s_0, so as to effectively use Stokes' formula. This we accomplish by adding to s_0 the harmonic function $t=t(\zeta, a)$ with singularities $-2 \log |\zeta-a|$ and $2 \log |\zeta-\zeta_0|$. For normalization we choose

$$t(\zeta, a)-2 \log |\zeta-\zeta_0| \to s_0(a) \tag{2}$$

as $\zeta \to \zeta_0$. The function

$$s(\zeta, a) = s_0(\zeta)+t(\zeta, a) \tag{3}$$

has a positive logarithmic pole at a as its only singularity. Thus $s(\zeta, a)$ qualifies as the proximity function for an arbitrary ζ on S.

The proximity function is symmetric:

$$s(a, b) = s(b, a) \tag{4}$$

for a, b in S. This is immediately seen by applying the Green's formula to $t(\zeta, a)$ and $t(\zeta, b)$ over S less small disks about a, b and ζ_0.

1C. In terms of t_0 we introduce on S the metric

$$d\omega = \lambda^2 \, dS, \tag{5}$$

where

$$\lambda^2 = \Delta s = \Delta s_0 = \frac{e^{t_0}|\text{grad}\, t_0|^2}{(1 + e^{t_0})^2} \tag{6}$$

and dS is the Euclidean area element in the parametric disk. It is easily seen that λ is finite everywhere and its only zeros are those of grad t_0.

In passing we note that, by virtue of $|\text{grad}\, t_0|^2 \, dS = dt_0 \, dt_0^*$, the total area of S is 4π:

$$\omega = \int_S d\omega = \int_{-\infty}^{\infty} \int_{\beta_x} \frac{e^{t_0}}{(1 + e^{t_0})^2} \, dt_0^* \, dx = 4\pi, \tag{7}$$

where $\beta_x = \{\zeta \mid t_0(\zeta) = x \in (-\infty, \infty)\}$.

The Gaussian curvature, defined at all points with $|\text{grad}\, t_0| \neq 0$, is constantly 1:

$$K = -\frac{\Delta \log \lambda}{\lambda^2} = 1. \tag{8}$$

2. The fundamental functions A, B, and C

2A. We turn to the domain surface R, an arbitrary Riemann surface. Take a parametric disk R_0 with boundary β_0 such that $f(\beta_0)$ does not meet $a_1, \cdots, a_q, \zeta_0, \zeta_1$. Let Ω be an adjacent regular region with boundary $\beta_0 \cup \beta_\Omega$. Denote by u the harmonic function in $\bar{\Omega}$ with $u \mid \beta_0 = 0$, $u \mid \beta_\Omega = k(\Omega) = \text{const.} > 0$ such that $\int_{\beta_0} du^* = 1$. The Dirichlet problem here and throughout the book is solvable, e.g., by the Perron method (see Ahlfors-Sario [1, p. 138]).

For $h \in [0, k]$ consider the level line $\beta_h = u^{-1}(h)$ and the region $\Omega_h = u^{-1}((0, h))$ bounded by $\beta_0 \cup \beta_h$. Without loss of generality we may assume that $f(\beta_0 \cup \beta_h)$ does not meet a. In fact, since the only singularity of s is logarithmic, the curvilinear integrals of s that we shall consider will be finite and continuous in h, and our formulas will extend to the case $a \in f(\beta_0 \cup \beta_h)$.

Denote by z_j the a-points of f in Ω_h and let $\alpha_j \subset \Omega_h$ be disjoint clockwise oriented level lines of $s(f(z), a)$ about z_j, encircling simply connected regions Δ_j. The curves β_h and β_0 are oriented so as to leave R_0 to the left. An application of Green's formula to $v(z)=h-u(z)$ and $s(f(z), a)$ gives

$$(9) \qquad \int_{\Sigma\alpha_j+\beta_h-\beta_0} v\,ds^*-s\,dv^* = \int_{\Omega_h-\cup\Delta_j} v\Delta_z s\,dR,$$

where dR is the Euclidean area element in the parametric disk. In view of $\int_{\alpha_j} dv^*=0$ we have $\int_{\alpha_j} s\,dv^* \to 0$ as α_j shrinks to z_j through level lines of s. Clearly $\int_{\alpha_j} ds^*$ tends to 4π times the multiplicity of the a-point z_j, whence

$$\int_{\Sigma\alpha_j} v\,ds^*-s\,dv^* \to 4\pi\sum v(z_j) = 4\pi\int_0^h (h-x)\,d\nu(x, a).$$

Here the sum is taken over all a-points in Ω_h, counted with their multiplicities, and $\nu(x, a)$ is the number of a-points in $\bar{R}_0 \cup \Omega_x$, counted similarly. We integrate $\int_0^h$ by parts and obtain from (9) in the limit,

$$(9)' \quad -4\pi h\nu(0, a)+4\pi\int_0^h \nu(h, a)\,dh+\int_{\beta_h-\beta_0} s\,du^*-h\int_{\beta_0} ds^* = \int_{\Omega_h} v(z)\,d\omega(f(z)).$$

2B. On the other hand, we consider the equation

$$(10) \qquad \int_{\beta_0+\Sigma\alpha_{0j}} ds^* = \int_{R_0-\cup\Delta_{0j}} \Delta_z s\,dR,$$

where the $\Delta_{0j} \subset R_0$ are small disks about the a-points of f in R_0. As the disks shrink to their centers, $\int_{\Sigma\alpha_{0j}} ds^*$ gives again $4\pi\nu(0, a)$ and we obtain

$$(10)' \qquad 4\pi\nu(0, a)+\int_{\beta_0} ds^* = \int_{R_0} d\omega(f(z)).$$

We multiply (10)′ by h and add to (9)′:

$$(11) \qquad 4\pi\int_0^h \nu(h, a)\,dh+\int_{\beta_h-\beta_0} s\,du^* = h\int_{R_0} d\omega+\int_{\Omega_h} v\,d\omega.$$

This is the preliminary form of Theorem 2E.

2C. Here the right-hand side has a simple meaning. To see this set $v=1$ in (9),

$$4\pi(\nu(h, a)-\nu(0, a))+\int_{\beta_h-\beta_0} ds^* = \int_{\Omega_h} d\omega,$$

which added to (10)′ gives

$$(12) \qquad 4\pi\nu(h, a)+\int_{\beta_h} ds^* = \int_{R_h} d\omega.$$

The left-hand side here is the h-derivative of the left-hand side of (11) except for a finite number of values of h in $[0, k]$ such that $a \in f(\beta_h)$. In fact, let $\Omega(\delta)$ be the part of R bounded by $\beta(\delta)=\beta_{h+\delta}-\beta_h$, the constant $\delta>0$ chosen so small that $\Omega(\delta) \cap f^{-1}(a)=\varnothing$. Then by Stokes' formula

$$\int_{\beta(\delta)} s\,du^* = \int_{\beta(\delta)} u\,ds^* - \int_{\Omega(\delta)} u\Delta_z s\,dR,$$

where

$$\begin{aligned}\int_{\beta(\delta)} u\,ds^* &= \delta\int_{\beta_{h+\delta}} ds^* + h\int_{\beta(\delta)} ds^* \\ &= \delta\int_{\beta_h} ds^* + \delta\int_{\Omega(\delta)} \Delta_z s\,dR + h\int_{\Omega(\delta)} \Delta_z s\,dR.\end{aligned}$$

It follows that

$$\begin{aligned}\left|\frac{1}{\delta}\int_{\beta(\delta)} s\,du^* - \int_{\beta_h} ds^*\right| &\leq \int_{\Omega(\delta)} |\Delta_z s|\,dR + \int_{\Omega(\delta)}\left|\frac{h-u}{\delta}\right| |\Delta_z s|\,dR \\ &\leq 2\int_{\Omega(\delta)} |\Delta_z s|\,dR,\end{aligned}$$

which tends to 0 with δ. The same is true for $\delta<0$, and we conclude that the right-hand side of (11) is the integral of $\int_{R_h} d\omega$.

2D. We are ready to introduce our fundamental quantities. The *A-function* (counting function)

$$A(h, a) = 4\pi\int_0^h \nu(h, a)\,dh \tag{13}$$

reflects the frequency of a-points. In particular, for a Picard point, i.e., a point not covered at all, $A \equiv 0$. The *B-function* (proximity function)

$$B(h, a) = \int_{\beta_h-\beta_0} s\,du^* \tag{14}$$

is the mean proximity to a of the image of $\beta_h-\beta_0$ under f. The *C-function* (characteristic function)

$$C(h) = \int_0^h\int_{R_h} d\omega\,dh \tag{15}$$

is the integral of the area of the (multisheeted) Riemannian image of $R_h = \bar{R}_0 \cup \Omega_h$ over S.

Our notations deviate from their counterparts in classical value distribution theory partly because of significantly different definitions, partly because our terminology conveys their meanings in a natural manner: the a-points are counted by the A-function, the boundary curves β are dealt with by the B-function, and the characteristic is the C-function.

2E. We have arrived at a generalization of Nevanlinna's [22] celebrated first main theorem on meromorphic functions in the plane (see 14B). Equality (11) takes on the following exact meaning, with no remainders to estimate (Sario [9]):

Theorem. *For complex analytic mappings of arbitrary Riemann surfaces into closed Riemann surfaces*

$$A(k, a) + B(k, a) = C(k). \tag{16}$$

This expresses a beautiful balance. If a point a is only lightly covered, then A is small but B must be correspondingly large; i.e., the image of $\beta_h - \beta_0$ comes close to the point a. Conversely, if a point a is strongly covered, then the image of $\beta_h - \beta_0$ stays at a great mean distance from a. In short, what a point a loses in its coverage, it gains in the proximity of the image of $\beta_h - \beta_0$. Moreover, the value of the sum $A + B$ is always C.

Note that the primary variable in (16) is not k but Ω which determines k.

We next ask: How many Picard points a_i, $i = 1, \cdots, q$, can there exist? To study this problem we must find a lower bound for $\sum A(k, a_i)$ or, what amounts to the same, an upper bound for $\sum B(k, a_i)$.

3. Euler characteristic

3A. As preparation we evaluate the Euler characteristic of Ω_h. Without loss of generality we may suppose that β_h consists of a finite number of analytic Jordan curves. This can indeed always be achieved by a sufficiently small increment of h. Denote by $\nu_h(\text{grad } u)$ the number of zeros of grad u in Ω_h.

Lemma. *The Euler characteristic* $e(h)$ *of* Ω_h *has the value*

$$\nu_h(\text{grad } u) = e(h). \tag{17}$$

The geometric meaning of this is clear from the following observation (Sario [5]). If Ω_h is doubly connected, the level lines of the harmonic conjugate u^* of u cover Ω_h smoothly without any branchings, and $\nu_h(\text{grad } u) = 0$. If, however, we have two contours constituting β_h, then some level line of u^* must branch off to reach both contours and we have a zero of grad u. In general, if the connectivity of Ω_h is c, the number of zeros of grad u is $c-2$.

Suppose then we have a "handle" in Ω_h, i.e., a torus-shaped part between β_0 and β_h. Then some level line of u^* must branch off before entering the tubes of the torus and again combine after completing its passage through the tubes. Thus we have two zeros of grad u, and if the genus is g we obtain $2g$ zeros by virtue of g handles.

In the general case of connectivity c and genus g we conclude that

$$\nu_h(\text{grad } u) = 2g+c-2.$$

This is the Euler characteristic of Ω_h.

The intuitive meaning of (17) is thus clear, and we proceed to give two proofs.

3B. The following demonstration making use of Riemann-Roch's theorem (see, e.g., Ahlfors-Sario [1, p. 324]) is due to Rodin [1].

We reflect $\bar{\Omega}_h$ about $\beta_0 \cup \beta_h$ so as to form its *double* $\hat{\Omega}_h$, a closed Riemann surface [*loc. cit.*, p. 119]. Let φ be the reflection of $\hat{\Omega}_h$, an indirectly conformal self-mapping leaving each point of $\beta_0 \cup \beta_h$ fixed. The genus $\hat{g}$ of $\hat{\Omega}_h$ is

$$\hat{g} = 2g+c-1,$$

where g is the genus and c the number of contours of Ω_h. Since u takes constant values on each component of $\beta_0 \cup \beta_h$, the differential ω defined by

$$\omega \,|\, \bar{\Omega}_h = du+i\,du^*,$$

$$\omega \,|\, \varphi(\bar{\Omega}_h) = -d(u\circ\varphi)-i\,d(u\circ\varphi)^*$$

is regular analytic on $\hat{\Omega}_h$. By Riemann-Roch's theorem ω has $2\hat{g}-2$ zeros in $\hat{\Omega}_h$, and consequently

$$\nu_h(\text{grad } u) = \nu_h(du+i\,du^*) = \hat{g}-1 = 2g+c-2 = e(h).$$

3C. A less function-theoretic but equally rapid proof (Sario [9]) can be given by making use of the well-known consequence of the Lefschetz fixed point theorem (see, e.g., Milnor [1, p. 37]): the sum of the indices of a differentiable vector field on a compact differentiable manifold is equal to its Euler characteristic.

Let $\hat{\Omega}_h$ and φ be as in 3B. The vector field X defined by

$$X \,|\, \bar{\Omega}_h = \text{grad } u,$$

$$X \,|\, \varphi(\bar{\Omega}_h) = -\text{grad } u\circ\varphi$$

is differentiable since u is constant on each component of $\beta_0 \cup \beta_h$. The sum of the indices of X is twice the number of zeros of grad u on Ω_h, counted with multiplicities. Therefore

$$2\nu_h(\text{grad } u) = \hat{e}(h),$$

where $\hat{e}(h)$ is the Euler characteristic of the closed surface $\hat{\Omega}_h$. Since

$$\hat{e}(h) = 2\hat{g}-2 = 2(2g+c-2) = 2e(h),$$

we conclude that (17) holds.

3D. As a corollary of Lemma 3A we have for the closed range surface S:

Lemma. *The number $\nu(\lambda)$ of zeros of λ in the metric* (5), (6) *is*

$$\nu(\lambda) = e_S+2 = 2g_S. \tag{18}$$

Here e_S is the Euler characteristic of S and g_S is its genus.

Let c be a positive number, so large that $\{\zeta \mid |t_0(\zeta)| > c\}$ consists of two disjoint "disks" about ζ_0 and ζ_1, respectively. Let $D_0 = \{\zeta \mid t_0(\zeta) > c\}$, $D_1 = \{\zeta \mid t_0(\zeta) < -c\}$ and $v = a(c-t_0)$, where a is a positive constant with $\int dv^* = 1$ along cycles of $B = S - \bar{D}_0 - \bar{D}_1$ separating D_0 and D_1. On applying Lemma 3A with B and v in place of Ω_h and u we obtain

$$\nu_B(\text{grad } v) = e(B).$$

Clearly

$$\nu_B(\text{grad } v) = \nu_B(\text{grad } t_0) = \nu(\text{grad } t_0)$$

and, by (6), $\nu(\text{grad } t_0) = \nu(\lambda)$. On the other hand, $e(B) = e_S+2$ and we arrive at (18).

3E. We can again give an alternate proof using the reasoning in 3C. With B and $t_0 \mid B$ in place of Ω_h and u we conclude that the number of zeros of grad t_0 is the Euler characteristic e_S+2 of B.

4. Areal proximity

4A. Our task of estimating $\sum B(h, a_i)$ is facilitated if B is replaced by the integral of its integral (cf. Introduction). The first integration means that the proximity to a of the image of the *curve* β_h is replaced by that of the *region* Ω_h. The second integration is for expediency.

For $i > 0$ and for any integrable φ defined in $[0, k]$ we set

$$\varphi_i(h) = \int_0^h \varphi_{i-1}(x)\, dx, \tag{19}$$

where φ_0 means φ. Our new proximity function is B_2 and subindex 2 can be appended to each term in (16).

Another simple device to shorten later reasoning is the following. Add to the points $a_1, \cdots, a_q$ the $2g$ zeros $a_{q+1}, \cdots, a_{q+2g}$ of λ, where g now stands for the genus of S, and for any function $\psi(h, a)$ set

$$\psi(h) = \sum_1^{q+2g} \psi(h, a_i). \tag{20}$$

Then

$$A_2(h) + B_2(h) = (q+2g)C_2(h). \tag{21}$$

We are to find an upper estimate for $B_2(h)$.

4B. On S we distribute a mass $dm = \sigma\, d\omega$ heavily concentrated at the points $a_1, \cdots, a_{q+2g}$. Specifically, we set

$$\sigma(\zeta) = \exp\Big[\sum_1^{q+2g} s(\zeta, a_i) - 2\log\Big(\sum_1^{q+2g} s(\zeta, a_i) + \text{const.}\Big)\Big],$$

where the constant is chosen to satisfy $\sum s(\zeta, a_i) + \text{const.} > 0$ and the logarithmic term serves to make the total mass $m = \int_S dm$ finite. In fact, for $r = |\zeta - a_i|$ in a parametric disk we have $\sigma(\zeta) = O(r^{-2}(\log r)^{-2})$, and the mass over $r < R$, say, is $O((\log R)^{-1}) = O(1)$. If an a_j, $j > q$, coincides with an a_i, $i \leq q$, then an obvious modification is needed. We can also choose ζ_0, ζ_1 in 1A so that this case does not occur.

The density $\sigma\lambda^2$ of dm induces in the $u + iu^*$-plane the density $\sigma\mu^2$, where

$$\mu(z) = \lambda(f(z))|f'(z)|\,|\text{grad}\, u(z)|^{-1}. \tag{22}$$

By the convexity property $\int_{\beta_h} \log\varphi\, du^* \leq \log \int_{\beta_h} \varphi\, du^*$ of the logarithm for any nonnegative function φ on β_h we have

$$B(h) < \int_{\beta_h} \log\sigma\, du^* + 2\log(B(h) + \text{const.}) + O(1),$$

$O(1)$ accounting for the $\int_{\beta_0}$ part of $B(h)$.

Decompose the integral into

$$\begin{aligned} F(h) &= \int_{\beta_h} \log(\sigma\mu^2)\, du^*, \\ G(h) &= -\int_{\beta_h} \log\mu^2\, du^*. \end{aligned} \tag{23}$$

Then

$$B_2(h) < F_2(h) + G_2(h) + 2[\log(C(h) + O(1))]_2 + O(h^2). \tag{24}$$

The purpose of the decomposition is that F_2 and G_2 can be estimated separately.

5. Main theorem

5A. We shall first find an upper bound for F_2. Set

$$H(h) = \int_{\beta_h} \sigma\mu^2\, du^*. \tag{25}$$

Clearly

$$F(h) \leq \log H(h),$$

while the inequality $\frac{1}{h}\int_0^h \log \psi(x)\,dx \leq \log\left(\frac{1}{h}\int_0^h \psi(x)\,dx\right)$ for nonnegative functions ψ in $(0, h)$ gives

$$F_1(h) \leq h \log\left(\frac{1}{h} H_1(h)\right) = h \log H_1(h) - h \log h.$$

Similarly

$$F_2(h) < h^2 \log H_2(h) + O(h^2 \log h). \tag{26}$$

To estimate H_2 note that

$$H_1(h) = \int_S (\nu(h, a) - \nu(0, a))\,dm(a) \leq \int_S \nu(h, a)\,dm(a).$$

Indeed, each of the first two expressions gives the total mass on the Riemannian image of Ω_h over S. On integrating (16) with respect to $dm(a)$ the first term thus has the lower bound $4\pi H_2(h)$. By (4), $s(\zeta, a) = s(a, \zeta)$, and from the construction of $s(\zeta, a)$ on the compact S it follows that the function is uniformly bounded from below. (In II.6 this will be proved on an arbitrary open or closed S.) We transpose $B(h, a)$ and conclude that $-B(h, a)$ on the right is dominated by $O(1)$:

$$4\pi H_2(h) < mC(h) + O(1).$$

Substitution into (26) gives

$$F_2(h) < h^2 \log\left(\frac{m}{4\pi} C(h) + O(1)\right) + O(h^2 \log h)$$

and therefore

$$F_2(h) < h^2(\log C(h) + O(\log h)). \tag{27}$$

5B. To compute $G_2(h)$ we first evaluate

$$G'(h) = -2\int_{\beta_h} d^* \log \mu.$$

Let Γ_j be small disks about the singularities of $\log \mu$, with $\partial\Gamma_j = \gamma_j$ oriented clockwise. Since $\log|f'(z)|$ and $\log|\text{grad}\, u(z)|$ are harmonic, we have $\Delta \log \mu = \Delta \log \lambda$ and consequently

$$G'(h) = 2\int_{\Sigma\gamma_j - \beta_0} d^* \log \mu - 2\int_{\Omega_h - \cup\Gamma_j} \Delta_z \log \lambda\, dR.$$

The last term has by (8), (5), (15) the value $2C'(h)+O(1)$. For any function φ defined in R_h we denote by $\nu(h, \varphi)$ the number of its zeros in R_h. As the Γ_j shrink to their centers we obtain by (22), (17)

$$G'(h) = 4\pi[-\nu(h, \lambda)-\nu(h, f')+e(h)]+2C'(h)+O(1),$$

where the zeros of $\lambda(f(z))$ and $f'(z)$ in R_0 are compensated for in $O(1)$. Triple integration gives

$$G_2(h) = -A_2(h, \lambda)-A_2(h, f')+E_2(h)+2C_2(h)+O(h^3) \tag{28}$$

with $E(h)=4\pi \int_0^h e(h)\, dh$.

5C. In the bracketed term of (24) we replace the integrand in both integrations by its value at the right end point of integration and obtain the estimate $2h^2 \log (C(h)+O(1))$. We substitute it, together with (27) and (28), into (24) and obtain the desired upper bound for $B_2(h)$. Substitution of this into (21) gives the main theorem (Sario [9]):

Theorem. *For an analytic mapping of an arbitrary Riemann surface R into a closed Riemann surface S*

$$(q+e_S)C_2(k) < \sum_1^q A_2(k, a_i)-A_2(k, f')+E_2(k)+O(k^3+k^2 \log C(k)). \tag{29}$$

In contrast with the classical theory, inequality (29) is valid for *all* regular subregions $\Omega \subset R$, with no "exceptional" ones omitted. Its meaning is that only relatively few points a_i can be sparsely covered. The sum of the corresponding A_2-functions (added to E_2) must dominate, in essence, $(q+e_S)C_2$. Explicit consequences of this result will be discussed in 7 to 9, and its relations with the Nevanlinna theory in 12C.

6. Nondegeneracy

6A. Significant conclusions can be drawn from (29) only if the mapping f is nondegenerate in the following sense:

$$\lim_{R_k \to R} \frac{k^3+k^2 \log C(k)}{C_2(k)} = 0. \tag{30}$$

We distinguish between four cases:

(a) k and $C(k)$ are bounded,

(b) k is unbounded, $C(k)$ is bounded,

(c) k is bounded, $C(k)$ is unbounded,

(d) k and $C(k)$ are unbounded.

By definition, R is *parabolic* or *hyperbolic* according as k is unbounded or bounded.

In case (a) inequality (29) gives no information; these mappings with bounded C-functions will be discussed in Chapter III. Case (b) cannot occur:

Lemma. *$C(k)$ is unbounded on every parabolic Riemann surface.*

Indeed, for any point a taken by f, $\nu(k, a) \geq 1$ from a certain R_k on, and $A(k, a) \geq 4\pi k + O(1)$.

Thus we only have to consider cases (c) and (d).

6B. The nature of condition (30) is clarified by considering the behavior of f at the ideal boundary β of R, i.e., the point $\beta \notin R$ in the Alexandroff compactification of R (see, e.g., Ahlfors-Sario [1, p. 8]). To avoid mappings which are of no interest, such as identity mappings, the classical value distribution theory only deals with functions that have at least one essential singularity. Mappings satisfying (30) are of this nature, since they possess an essential singularity on β in the following sense:

Lemma. *Condition* (30) *implies that f comes arbitrarily close to any point a in every boundary neighborhood of R.*

By a boundary neighborhood we mean the complement of a compact set.

Proof. If the lemma were false, there would exist an $M < \infty$, a point $a \in S$ and a compact set $K \subset R$ such that $s(f(z), a) < M$ for all $z \in R - K$. This would imply that $B(k, a) = O(1)$ for $R_k \supset K$, and $A(k, a) \leq 4\pi\nu_K(a)k$, where $\nu_K(a)$ is the number of a-points in K. Consequently $C(k) = O(k)$, $\log C(k) = O(\log k)$, $C_2(k) = O(k^3)$, and $(k^3 + k^2 \log C(k))/C_2(k)$ does not tend to 0.

We shall return to these questions in IV.6–7.

6C. Condition (30) can be given a simpler form (Rao):

Lemma. *The mapping f is nondegenerate in the sense of* (30) *if there exists a constant α, $0 < \alpha < 1$, such that*

$$\lim_{R_k \to R} \frac{k}{C(k)} = 0 \quad \textit{and} \quad \lim_{R_k \to R} \frac{\log C(k)}{C(\alpha k)} = 0. \tag{31}$$

Proof. The first condition implies that $C(k) \to \infty$ as $R_k \to R$, and it is not difficult to show that this in turn gives $k^3/C_2(k) \to 0$. On the other hand, as a consequence of the second condition (31) we have

$$k^2 \log C(k)/C_2(k) \to 0.$$

In fact,

$$C_1(h) = \int_0^h C(t)\,dt \geq \int_{\sqrt{\alpha}\,h}^h C(t)\,dt \geq h(1-\sqrt{\alpha})C(\sqrt{\alpha}h),$$

and therefore

$$C_2(k) \geq \int_0^k t(1-\sqrt{\alpha})C(\sqrt{\alpha}t)\,dt$$

$$\geq \int_{\sqrt{\alpha}k}^k t(1-\sqrt{\alpha})C(\sqrt{\alpha}t)\,dt \geq (1-\sqrt{\alpha})C(\alpha k)k^2\,\frac{1-\alpha}{2}.$$

This proves our last claim and establishes that (30) indeed follows from (31).

7. Exceptional points

7A. We introduce what we shall call the *Euler index*

$$\eta = \liminf_{R_k \to R} \frac{E_2(k)}{C_2(k)} \tag{32}$$

and obtain the following extension of Picard's theorem:

Theorem. *For nondegenerate mappings of arbitrary Riemann surfaces into closed Riemann surfaces the number P of Picard points has the bound*

$$P \leq \eta - e_S. \tag{33}$$

If R is a sphere less a finite number of points, then $P \leq -e_S$. This elegant result was first obtained by Chern [1]. It remains true for all cases with bounded $e(k)$ and, more generally, with $\eta = 0$.

7B. The following question now arises: What about values a which, although not completely uncovered, have the Picard-Borel property of being so sparsely covered that $A_2(k, a)/C_2(k) \to 0$? The answer is given by the following extension to Riemann surfaces of the Picard-Borel theorem:

There exist at most $\eta - e_S$ points with the Picard-Borel property.

The proof is immediate: the substitution $q = \eta - e_S + \epsilon$ with $\epsilon > 0$ gives the contradiction $\eta + \epsilon \leq \eta$.

7C. Suppose then a is even more strongly covered, but not necessarily with the complete power that makes $A_2(k, a)/C_2(k) \to 1$. We introduce the *defect*

$$\alpha(a) = 1 - \limsup_{R_k \to R} \frac{A_2(k, a)}{C_2(k)} = \liminf_{R_k \to R} \frac{B_2(k, a)}{C_2(k)}. \tag{34}$$

The question here is: What can be said about the sum of all defects?

Defect relation. *The defect sum has the bound*

$$\sum \alpha(a) \leq \eta - e_S. \tag{35}$$

Thus the defective points need not be located exclusively at the $\leq \eta - e_S$ Picard points or the $\leq \eta - e_S$ Picard-Borel points, but may be spread to a countable point set provided the sum of the defects does not exceed $\eta - e_S$. If $\eta < \infty$, then it is easily seen that the set of defect points is indeed countable and the defect relation is a direct consequence of Theorem 5C.

8. Ramification

8A. The above applications concern defective coverage of a point a. We now ask about multiple coverage, i.e., coverage by algebraic branch points. What can be said about the set of points that are strongly covered by multiple points? (For basic concepts on covering surfaces see, e.g., Ahlfors-Sario [1].)

We introduce the *ramification index*

$$\beta(a) = \liminf_{R_k \to R} \frac{A_2(k, f_a')}{C_2(k)}, \tag{36}$$

where $A(k, f_a')$ counts the sum of the orders of branch points above a. Since $f(R_k)$ only has a finite number of branch points it is meaningful to take the *total ramification*

$$\sum \beta(a) \leq \liminf_{R_k \to R} \frac{A_2(k, f')}{C_2(k)}, \tag{37}$$

where the sum is extended over all points a of S.

Ramification relation. *The total ramification has the bound*

$$\sum \beta(a) \leq \eta - e_S. \tag{38}$$

8B. The defect and ramification relations are, of course, special cases of

$$\sum \alpha(a) + \sum \beta(a) \leq \eta - e_S. \tag{39}$$

For the sum $\alpha(a) + \beta(a)$ with given a we have the inequality

$$\alpha(a) + \beta(a) \leq 1. \tag{40}$$

This is obtained by dividing the twice integrated (16) by $C_2(k)$ and by observing that $A_2(k, f_a') \leq A_2(k, a)$.

8C. The contribution to $C_2(k)$ of the sum $A_2(k, f_a') + B_2(k, a)$ is measured by

$$\sigma(a) = \liminf_{R_k \to R} \frac{A_2(k, f_a') + B_2(k, a)}{C_2(k)}. \tag{41}$$

The meaning of $\sigma(a)$ is made clear by considering the number $\bar{\nu}(k, a)$ of a-points of f, each counted only once. We set $\bar{A}(h, a) = 4\pi \int_0^h \bar{\nu}(h, a)\, dh$ and note that $\bar{\nu}(k, a) = \nu(k, a) - \nu(k, f_a')$ and $A(k, a) = \bar{A}(k, a) + A(k, f_a')$. It follows that

$$A_2(k, f_a') + B_2(k, a) = C_2(k) - \bar{A}_2(k, a)$$

and consequently

$$\sigma(a) = 1 - \limsup_{R_k \to R} \frac{\bar{A}_2(k, a)}{C_2(k)}. \tag{42}$$

For the sum of the $\sigma(a)$ we have the bound

$$\sum \sigma(a) \leq \eta - e_S. \tag{43}$$

In fact,

$$\sum \liminf_{R_k \to R} \frac{A_2(k, f_a') + B_2(k, a)}{C_2(k)} \leq \liminf_{R_k \to R} \frac{A_2(k, f') + \sum B_2(k, a)}{C_2(k)} \leq \eta - e_S.$$

A point a is called *totally ramified* if the equation $f(z) = a$ has no simple roots.

Theorem. *The number of totally ramified points does not exceed* $2(\eta - e_S)$.

Indeed, for such a, $\nu(k, a) \geq 2\bar{\nu}(k, a)$ and one concludes that

$$\sigma(a) \geq 1 - \frac{1}{2} \limsup_{R_k \to R} \frac{A_2(k, a)}{C_2(k)} \geq \frac{1}{2}.$$

The statement follows from (43).

§2. MEROMORPHIC FUNCTIONS ON ARBITRARY RIEMANN SURFACES

In the special case where the range surface S is the Riemann sphere our mappings f are meromorphic functions on an arbitrary Riemann surface R, and the fundamental B- and C-functions will take on concrete meanings. We obtain the bound $2 + \eta$ for the number of Picard values and show that this bound is sharp. For surfaces R_p that carry capacity functions with compact level lines we obtain generalizations and new proofs of Nevanlinna's classical defect and ramification relations, without using exceptional intervals.

9. Main theorems

9A. We again denote the local parameter of S and the generic variable in the plane by the same letter ζ and, using the notations of §1, choose $\zeta_0 = \infty$, $\zeta_1 = 0$. Then

$$t_0(\zeta) = 2 \log |\zeta|,$$

and

(44) $$s(\zeta, a) = \log \frac{(1+|\zeta|^2)(1+|a|^2)}{|\zeta-a|^2} = 2 \log \frac{1}{[\zeta, a]},$$

where $[\zeta, a]$ is the chordal distance between the stereographic images of ζ and a on the Riemann sphere (of diameter 1). Thus the B-function is twice the mean logarithmic chordal proximity to a of the image of $\beta_h - \beta_0$.

The mass element

(45) $$d\omega = \frac{4}{(1+|\zeta|^2)^2} dS,$$

with dS the Euclidean area element, is the (4-fold) area element of the Riemann sphere, and the derivative of the C-function is the (4-fold) spherical area of the Riemannian image of R_h.

These meanings of B and C add to the content of the first main theorem,

(46) $$A(k, a) + B(k, a) = C(k),$$

in the present case.

9B. The second main theorem reads:

Theorem. *For meromorphic functions on arbitrary Riemann surfaces*

(47) $$(q-2)C_2(k) < \sum_1^q A_2(k, a_i) - A_2(k, f') + E_2(k) + O(k^3 + k^2 \log C(k)).$$

Again, this result holds for all subregions Ω, with no exceptional ones omitted (cf. 18A). For a more detailed discussion of this aspect we refer the reader to the Introduction.

For functions satisfying the nondegeneracy condition (30) on an arbitrary R we now have the essential requirement reflecting the topology of R: the C-function must grow at least as rapidly as the Euler characteristic. For these mappings we have obtained a generalized Picard theorem:

The number of Picard values cannot exceed the excess of η over -2,

(48) $$P \leq 2 + \eta.$$

More accurately, (47) yields

(49) $$P \leq 2 + \eta - \liminf_{R_k \to R} \frac{A_2(k, f')}{C_2(k)}.$$

Particularly illustrative of this general theory is the Gaussian mapping of minimal surfaces (App. II).

10. Sharpness of even bounds

10A. In reaching the universal bound $2+\eta$ we were rather generous in suppressing various terms in deriving the second main theorem, and one might wonder whether the bound is at all sharp. We shall show that the bound is the best possible in the sense that, for any integer $\eta \geq 0$ there exists a Riemann surface R and a meromorphic function f on R such that $P=2+\eta$.

We start with an example giving the proof for even η (Sario [5]). Consider a covering surface R of the z-plane that consists of n sheets with branch points of multiplicity n at $z=i\pi(\frac{1}{2}+m)$, $m=0, \pm 1, \pm 2, \cdots$. The sheets are attached to each other in the usual manner along the edges of the slits from $z=i\pi(\frac{1}{2}+2m)$ to $i\pi(\frac{3}{2}+2m)$. We denote the projection map by the same symbol z as the local parameter. The "capacity function" (cf. 12A)

$$p(z) = \frac{1}{2\pi n} \log |z| \tag{50}$$

serves to give the exhausting regions $R_h = \{z \mid |z| < e^{2\pi n h}\}$. Moreover, the restriction of p to any $\bar{\Omega}_h = \bar{R}_h - R_0$ is our function u with flux 1. We consider the well-known function

$$f(z) = \sqrt[n]{\frac{e^z+i}{e^z-i}}, \tag{51}$$

meromorphic on R.

10B. The Euler characteristic $e(G)$ of a triangulated surface G is, by definition, $-n_0+n_1-n_2$, with n_0, n_1, n_2 the numbers of vertices, edges, and faces. We make use of the Hurwitz formula

$$e(R_h) = ne(R_h') + \sum b,$$

where R_h' is the projection of R_h, and $\sum b$ is the sum of the orders of branch points of R_h. The formula is immediately verified by taking a triangulation of R_h' that includes among the vertices the projections of the branch points. On lifting the triangulation to R_h the numbers n_1 and n_2 are multiplied by the number n of sheets, while the number of vertices of R_h falls short of nn_0 by $\sum b$.

10C. Since $e(R_h') = -1$ and $\sum b$ above $|z| < 2\pi$ is $4(n-1)$, we obtain on disregarding bounded quantities,

$$e(h) = e(\Omega_h) \sim 4(n-1) \frac{e^{2\pi n h}}{2\pi}.$$

Consequently

$$E(h) \sim \frac{4(n-1)}{\pi n} e^{2\pi nh}$$

and

(52)
$$E_2(k) \sim \frac{n-1}{\pi^3 n^3} e^{2\pi nk}.$$

The poles of f are the zeros of $e^z - i$, i.e., $z_j = i(\pi/2 + 2\pi j)$ with all integers j, and it follows that

$$\nu(h, \infty) \sim 2 \frac{e^{2\pi nh}}{2\pi}.$$

A fortiori

$$A(h, \infty) \sim \frac{2}{\pi n} e^{2\pi nh}$$

and

(53)
$$A_2(k, \infty) \sim \frac{1}{2\pi^3 n^3} e^{2\pi nk}.$$

We therefore have, by virtue of the uniform boundedness from below of $s(\zeta, a)$,

(54)
$$\eta = \liminf_{k\to\infty} \frac{E_2(k)}{C_2(k)} \leq \liminf_{k\to\infty} \frac{E_2(k)}{A_2(k, \infty)} = 2n-2.$$

Since the nondegeneracy condition (30) is obviously satisfied, (48) states that we can have no more than $2n$ Picard values. But this is precisely the number of values $e^{i\mu\pi/n}$, $\mu = 0, \cdots, 2n-1$, not taken by f.

10D. The conformal mapping effected by f is simple and illuminating. We first take the n-sheeted horizontal strip H of R between $y=0$ and $y=2\pi$. It is mapped by $s=e^z$ onto the n-sheeted s-plane H_s slit along the positive real axis and with branch points of order $n-1$ at $s=\pm i$. The linear function $t=(s+i)/(s-i)$ maps H_s onto an n-sheeted t-plane H_t slit along the upper half of $|t| = 1$, and with branch points of order $n-1$ at $t=0, \infty$. The function $\zeta = \sqrt[n]{t}$ maps H_t onto a 1-sheeted region H_ζ^0 of the ζ-plane less slits L_μ along $|\zeta|=1$ from $e^{2\mu\pi i/n}$ to $e^{(2\mu+1)\pi i/n}$, $\mu=0, \cdots, n-1$.

Each n-sheeted strip $2\pi m < y < 2\pi(m+1)$ of R is mapped by f onto a duplicate H_ζ^m of H_ζ^0. The image R_ζ of R is obtained by identifying the inner edges $|\zeta| = 1-0$ of all slits L_μ on H_ζ^m with the outer edges $|\zeta|=1+0$ of the corresponding slits L_μ on H_ζ^{m+1}. The process creates logarithmic branch points at the end points of the slits L_μ. These are the $2n$ Picard points of f.

11. Sharpness of arbitrary bounds

11A. The surface described above has no algebraic branch points, hence $A(k, f') \equiv 0$. The question now arises whether or not the bound in (49) can be reached for an arbitrary positive integer η when the surface is so strongly ramified that lim inf $(A_2(k, f')/C_2(k)) > 0$. We again use (51) and form the function $\hat{f} = \zeta^N$, where N is a factor of $2n$. Similar considerations as in 10C give for $N > 1$

$$\liminf_{k \to \infty} \frac{A_2(k, \hat{f}')}{e^{2\pi nk}} > 0.$$

From the mapping property described in 10D it follows that

$$\limsup_{k \to \infty} \frac{C_2(k, \hat{f})}{e^{2\pi nk}} > 0.$$

Therefore

$$\tau = \liminf_{k \to \infty} \frac{A_2(k, \hat{f}')}{C_2(k, \hat{f})} > 0.$$

A computation analogous to the one in 10C yields $\eta \leq (2n-2)/N$, and we conclude by (49) that

$$P \leq \frac{2n}{N} + 2\left(1 - \frac{1}{N}\right) - \tau.$$

For $N = 1$, $\tau = 0$ and we again have the bound $2n$. Clearly the number of Picard values of $\hat{f}$ is $2n/N$. For any integer $q \geq 2$ we can choose $n = q$ and $N = 2$, say, and obtain q Picard values. In this case (49) becomes

$$P \leq q + (1 - \tau)$$

and since P and q are integers, P cannot exceed q. We have shown that the bound in (49) is sharp for all positive integers.

11B. We return to the question of the sharpness of (48). The following interesting modification of the surface of 10A was given by Rodin [2] and shows the sharpness for every integer η.

Take n copies of the finite z-plane $z = x + iy$, each slit along the rays

$$I_m = \{z \mid x \leq 0, y = 2\pi i m\}, \tag{55}$$

$m = 0, \pm 1, \pm 2, \cdots$. For a fixed m the edges of I_m on all the sheets are identified so as to obtain a branch point of multiplicity n at $2\pi i m$. This describes a covering surface R of the plane. We again choose $p(z) = (2\pi n)^{-1} \log |z|$ and define the region R_h as before.

On this surface we consider the meromorphic function

$$f(z) = \sqrt[n]{\frac{e^{z+\pi i}}{e^{z+\pi i}+1}} \tag{56}$$

with poles at $z = 2\pi im$. We have

$$\nu(h, \infty) \sim 2\,\frac{e^{2\pi nh}}{2\pi}$$

and

$$A_2(k, \infty) \sim \frac{1}{2\pi^3 n^3}\,e^{2\pi nk}. \tag{57}$$

In view of the branch points of R over the points $2\pi im$ we obtain

$$e(h) \sim 2(n-1)\,\frac{e^{2\pi nh}}{2\pi},$$

$$E(h) \sim \frac{2(n-1)}{\pi n}\,e^{2\pi nh}, \tag{58}$$

$$\eta \leq \liminf_{k\to\infty} \frac{E_2(k)}{A_2(k, \infty)} = n-1, \tag{59}$$

and the bound is $2+\eta \leq n+1$.

On the other hand, the function f omits the values 0 and $e^{2\pi i\mu/n}$, $\mu = 0, \cdots, n-1$. Thus $P = n+1$, which establishes the assertion.

12. The class of R_p-surfaces

12A. In the examples of 11 we were able to replace the directed limits of the general theory by ordinary limits. This is possible whenever R is an R_p-*surface* characterized by the following condition: there exists a function p harmonic on R with a finite number of negative logarithmic poles and with compact level lines. Then R_0 can be chosen as the region bounded by some level line β_0: $p = c$, and u on any Ω bounded by β_0 and another level line β_Ω is the restriction to Ω of the *same* function $p - c$.

The function (50) is a special case of a *capacity function*, constructed on an arbitrary Riemann surface as follows. (The construction is the same for one or several singularities.)

For any set E let $H(E)$ be the space of harmonic functions on E. Given a Riemann surface R choose a point $z_0 \in R$, a parametric disk D containing z_0, and a regular subregion Ω of R containing $\bar{D}$. The capacity function p_Ω of Ω with singularity at z_0 is, by definition, the function $p_\Omega \in H(\bar{\Omega} - z_0)$ with

$$p_\Omega \,|\, D = \frac{1}{2\pi} \log |z - z_0| + h(z), \tag{60}$$

where $h \in H(D)$, $h(z_0)=0$, and $p_\Omega \mid \partial\Omega = k_\Omega = \text{const}$. For $\Omega \subset \Omega'$ it is seen that $k_\Omega \le k_{\Omega'}$ (App. I.§1), and the directed limit

$$k_\beta = \lim_{\Omega \to R} k_\Omega \tag{61}$$

exists. Without using the monotonicity of k_Ω we can also define k_β as $\sup k_\Omega$. This gives the capacity

$$c_\beta = e^{-k_\beta} \tag{62}$$

of the ideal boundary β of R (Sario [3]). The surface is *parabolic* or *hyperbolic* according as $c_\beta = 0$ or $c_\beta > 0$. The class of parabolic surfaces is denoted by O_G and can be shown to coincide with the class of Riemann surfaces on which there are no Green's functions (App. I.§1).

12B. On an arbitrary $R \notin O_G$ the directed limit

$$p_\beta = \lim_{\Omega \to R} p_\Omega \tag{63}$$

exists [*loc. cit.*] and is, by definition, the capacity function p_β on R. However, there clearly are hyperbolic surfaces on which the level lines of p_β are not all compact. E.g., on the unit disk punctured at $z=1/2$, $p_\beta = (1/2\pi)\log|z|$ has a noncompact level line $|z|=1/2$.

On a parabolic surface R the directed limit (63) does not always exist, but on every R there is a nested exhausting sequence $\{\Omega_n\}$ such that the corresponding capacity functions p_n converge [*loc. cit.*]. The limit p_β is again defined as a capacity function on R. In contrast with the Green's function there thus exists a capacity function on every Riemann surface, although for $R \in O_G$ it may not be unique. The usefulness of p_β in this case is in that every $R \in O_G$ possesses a capacity function with compact level lines and is thus an R_p-surface. This is a recent result of Nakai [1], to be proved in Chapter IV.

Other aspects of R_p-surfaces will be discussed in VI.19.

12C. For surfaces R_p that possess capacity functions with compact level lines we can replace A_2, B_2, C_2, E_2 by A, B, C, E in defining α, β, η. In fact, if we set

$$\delta(a) = \liminf_{k \to k_\beta} \frac{B(k, a)}{C(k)}, \tag{64}$$

$$\vartheta(a) = \liminf_{k \to k_\beta} \frac{A(k, f_a')}{C(k)}, \tag{65}$$

then by the reasoning leading to l'Hospital's rule, $\delta \le \alpha$, and $\vartheta \le \beta$. Similarly we write

$$\kappa = \liminf_{k \to k_\beta} \frac{E(k)}{C(k)} \tag{66}$$

and obtain *the defect and ramification relation in the general case of mappings into closed surfaces*:

$$\sum \delta(a) + \sum \vartheta(a) \leq \kappa - e_S. \tag{67}$$

For meromorphic functions on Riemann surfaces the bound is $2+\kappa$.

In the special case of meromorphic functions in the disk $|z| < R \leq \infty$ this reduces to Nevanlinna's classical defect and ramification relation

$$\sum \delta(a) + \sum \vartheta(a) \leq 2, \tag{68}$$

which we have thus obtained from the second main theorem without exceptional intervals, and simultaneously for the plane and the disk.

§3. SURFACES R_s AND CONFORMAL METRICS

For the sake of completeness and comparison we shall also derive the defect and ramification relation using a more classical approach that involves exceptional intervals. Several aspects in our reasoning go back to the fundamental work of Ahlfors [10] on meromorphic functions in the plane. To develop a theory on certain Riemann surfaces R_s we make use of a metric to exhaust R_s. The metric we choose is suggested by $ds = |\text{grad } p_\beta| \, |dz|$, where p_β is the capacity function with compact level lines, as exemplified in 11 and 12. We shall, however, consider a slightly more general situation, which offers interest in its own right.

Our reasoning is self-contained and independent of the preceding sections.

13. Metric

13A. Let R be a Riemann surface endowed with a conformal metric

$$ds = \lambda(z)|dz|, \tag{69}$$

where $\lambda \geq 0$ is defined in each parametric disk and ds is invariant under change of parameter. The length $l(\alpha)$ of a rectifiable arc α on R is well defined, and the distance $d(z_1, z_2)$ between two points is $\inf l(\alpha)$ for arcs α from z_1 to z_2. The distance $d(E_1, E_2)$ between two subsets of R is defined as $\inf d(z_1, z_2)$ for $z_1 \in E_1$, $z_2 \in E_2$.

Let R_0 be a regular subregion with boundary β_0. For $\sigma > 0$ let

$$\beta_\sigma = \{z \in R \mid d(z, R_0) = \sigma\}. \tag{70}$$

We impose the following requirements on our metric:

(a) Log λ is harmonic except for logarithmic singularities.

(b) The length of the "level line" β_σ is constantly

$$\int_{\beta_\sigma} ds = 1. \tag{71}$$

(c) The level lines β_σ are compact for every $\sigma > 0$.

The last condition is understood to mean that β_σ is either compact or void. We set $\sigma_\beta = \sup \sigma$ for nonvoid β_σ, and distinguish between $\sigma_\beta = \infty$ and $\sigma_\beta < \infty$. In the special case $\lambda = |\text{grad } p_\beta|$ this distinction is equivalent with that of parabolic and hyperbolic surfaces.

A Riemann surface with a metric ds satisfying conditions (a) to (c) shall be called an R_s-surface. We shall not discuss here the interesting differential-geometric problem of characterizing all R_s-surfaces.

13B. Schematically the parameter σ and the arc length s along β_σ constitute a coordinate system on $R - \bar{R}_0$. If R_σ is the relatively compact region bounded by β_σ, then $\Omega_\sigma = R_\sigma - \bar{R}_0$ corresponds to a rectangle with width σ and height 1 in the (σ, s)-plane. A concrete illustration is given by $\lambda = |\text{grad } u|$ for a harmonic function u on $\bar{R}_\sigma - R_0$ with $u \mid \beta_0 = 0$, $u \mid \beta_\sigma =$ const. $= \sigma$ such that $\int_{\beta_0} du^* = 1$. For genus $g \geq 0$ and connectivity $c \geq 2$ of Ω_σ, $2g + 2(c-2)$ horizontal slits appear in the (σ, s)-rectangle; the edges of the slits and the horizontal sides of the rectangle are suitably identified to form a conformal equivalent of Ω_σ. The slits issue from the zeros of $|\text{grad } u|$. The g "handles" of Ω_σ give rise to $2g$ slits in the interior of the rectangle, and the c contours produce $2(c-2)$ slits terminating at the vertical edges of the rectangle. In the general case $ds = \lambda |dz|$ the end points of the slits are at the zeros of λ. The rate of growth of the number of these zeros will play a fundamental role in our approach.

14. The fundamental functions

14A. Let $\zeta = f(z)$ be a meromorphic function on an R_s-surface and denote by $\nu(\sigma, a)$ the number of its a-points, counted with multiplicities, in R_σ. The A-function is defined as

$$A(\sigma, a) = 4\pi \int_0^\sigma \nu(\sigma, a)\, d\sigma, \tag{72}$$

the B-function as

$$B(\sigma, a) = 2 \int_{\beta_\sigma - \beta_0} \log \frac{1}{[f(z), a]}\, ds, \tag{73}$$

and the C-function as

$$C(\sigma) = A(\sigma, \infty) + B(\sigma, \infty). \tag{74}$$

Differentiation of (73) gives for any a, b, finite or infinite, and for a σ such that λ has no zeros on β_σ,

$$\begin{aligned} \frac{dB(\sigma, a)}{d\sigma} - \frac{dB(\sigma, b)}{d\sigma} &= 2\int_{\beta_\sigma} \frac{d}{d\sigma} \log \left| \frac{\zeta - b}{\zeta - a} \right| ds \\ &= 2\int_{\beta_\sigma} d \arg \frac{\zeta - b}{\zeta - a} = 4\pi(\nu(\sigma, b) - \nu(\sigma, a)), \end{aligned} \tag{75}$$

where $d/d\sigma$ stands for the exterior normal derivative in the metric under consideration. The differentiation under the integral sign is legitimate, for $\int_{\beta_\sigma}$ is an integral with respect to s from 0 to 1. On integrating (75) from 0 to σ we again obtain the first main theorem (46) for the present special case:

$$A(\sigma, a) + B(\sigma, a) = C(\sigma). \tag{76}$$

14B. We observe in passing that the theorem can of course also be written in the classical form. Let

$$m(\sigma, a) = \frac{1}{2\pi} \int_{\beta_\sigma} \log \frac{1}{[f(z), a]} ds$$

be the "proximity function", $N(\sigma, a) = \int_0^h \nu(\sigma, a)\, d\sigma$ the "counting function", and

$$T(\sigma) = m(\sigma, \infty) + N(\sigma, \infty)$$

the "characteristic function". Then

$$N(\sigma, a) + m(\sigma, a) = T(\sigma) + O(1).$$

In the case of the z-plane R and for $ds = |dz|/2\pi r$ this is Nevanlinna's [22] first main theorem. The remainder $O(1)$ is bounded for each a, but the bound varies with a.

14C. As is to be expected, the Shimizu [2]-Ahlfors [1] interpretation of the C-function continues to be valid in the present case. In integrating (76) over the area elements $d\alpha(a)$ of the a-sphere S the integral of $\log [\zeta, a]^{-1}$ is independent of ζ, and the integral of $B(\sigma, a)$ vanishes. We obtain

$$C(\sigma) = \frac{1}{\pi} \int_S A(\sigma, a)\, d\alpha(a) = 4 \int_S \int_0^\sigma \nu(\sigma, a)\, d\sigma\, d\alpha(a). \tag{77}$$

For convenience we shall indicate differentiation by subindices and use the notation

$$|f_s| = \frac{|df/dz|}{|ds/dz|} = |f_z|\lambda^{-1}. \tag{78}$$

Then the spherical area of the Riemannian image of R_σ under f is

$$S(\sigma) = \int_S \nu(\sigma, a)\, d\alpha(a) = \int_0^\sigma d\sigma \int_{\beta_\sigma} \frac{|f_s(z)|^2}{(1+|f(z)|^2)^2}\, ds, \tag{79}$$

where the last expression is in anticipation of later reasoning. We conclude that

$$C(\sigma) = 4\int_0^\sigma S(\sigma)\, d\sigma. \tag{80}$$

The derivative of the C-function is the (4-fold) spherical area $4S(\sigma)$.

As a corollary we see that $C(\sigma)$ is convex in σ.

15. Preliminary form of the second main theorem

15A. Our next question is: What are the comparative contributions of A and B to C? In particular, for how large a set of points a can B make any substantial contributions? The answer will be given by the second main theorem which we shall now prove. It will give us an upper bound for the sum of the B-functions for any finite number of points a.

We use a mass distribution

$$d\mu(a) = \rho(a)\, d\alpha(a) \tag{81}$$

with density $\rho(a)$ and total mass 1 on the sphere S with diameter 1 above the ζ-plane. We simply postulate that $\rho(a)$ is sufficiently regular to justify subsequent operations on it. This condition will be obviously met by the particular ρ we shall use.

We consider our rectangle in the (σ, s)-plane, our region Ω_σ, the ζ-plane, and the S-sphere. The area element $d\alpha(a)$ is the image of some area element at the point z of Ω_σ under the function f and the subsequent stereographic projection. The area element at z also has an image in our (σ, s)-rectangle, and we consider the induced density $\delta(z)$ at this image. Clearly it is equal to the corresponding density $\rho(f(z))$ multiplied by the change of scale of the area element:

$$\delta(z) = \frac{|f_s(z)|^2}{(1+|f(z)|^2)^2}\, \rho(f(z)). \tag{82}$$

By virtue of the convexity of the logarithm we have

$$\int_{\beta_\sigma} \log \delta \, ds \leq \log \int_{\beta_\sigma} \delta \, ds$$

or, equivalently,

$$\int_{\beta_\sigma} \log \frac{\delta}{\rho} \, ds + \int_{\beta_\sigma} \log \rho \, ds \leq \log \int_{\beta_\sigma} \delta \, ds, \tag{83}$$

where ρ stands for $\rho(f(z))$.

This is the preliminary form of the second main theorem. The right-hand side can be shown to be an insignificant remainder term. The second term on the left depends only on the mass distribution. If the mass is suitably concentrated at q given points a_i, the term will give, in essence, the sum of the B-functions for these points. But that is precisely what we want to estimate. The first term depends only on f and λ. It will be evaluated making use of the C-function, the A-function for the multiple points of f, and the A-function for the zeros of λ. Thus the sum $\sum B(\sigma, a_i)$ will be estimated in terms of $C(\sigma)$, $A(\sigma, f')$, and $A(\sigma, \lambda)$. This is the second main theorem.

16. Evaluations

16A. We start with the first term in (83) and evaluate

$$K(\sigma) = \int_{\beta_\sigma} \log \frac{\delta}{\rho} \, ds = 2 \int_{\beta_\sigma} \log \frac{|f_s|}{1+|f|^2} \, ds. \tag{84}$$

We form the derivative

$$K'(\sigma) = 2 \frac{d}{d\sigma} \int_{\beta_\sigma} \log \frac{1}{1+|f|^2} \, ds + 2 \int_{\beta_\sigma} \frac{d}{d\sigma} \log |f_z \lambda^{-1}| \, ds. \tag{85}$$

Denote the first and second terms on the right by I_1 and I_2. To evaluate I_1 we have from (73)

$$B(\sigma, \infty) = -\int_{\beta_\sigma - \beta_0} \log \frac{1}{1+|f|^2} \, ds \tag{86}$$

and consequently

$$I_1 = -2 \frac{dB(\sigma, \infty)}{d\sigma}. \tag{87}$$

Since $\log \lambda$ is harmonic except for logarithmic poles the argument principle can be applied to I_2:

$$I_2 = 4\pi[\nu(\sigma, f_z) - \nu(\sigma, f_z^{-1}) - \nu(\sigma, \lambda)], \tag{88}$$

where in general $\nu(\sigma, \varphi)$ stands for the number of zeros, counted with multiplicities, of any function φ defined in R_σ. The number $\nu(\sigma, f')$ of multiple points of f in R_σ, each point of multiplicity k counted $k-1$ times, is

(89) $$\nu(\sigma, f') = \nu(\sigma, f_z) - \nu(\sigma, f_z^{-1}) + 2\nu(\sigma, f^{-1}).$$

Indeed, a finite value of f with multiplicity k is a zero of f_z of multiplicity $k-1$. As to a pole of f of multiplicity k, it is a pole of f_z of multiplicity $k+1$, and the second term on the right gives us $-k-1$. But the third term contributes $2k$, and the total of the last two terms is again $k-1$ as desired. With this notation I_2 becomes

(90) $$I_2 = 4\pi[\nu(\sigma, f') - 2\nu(\sigma, f^{-1}) - \nu(\sigma, \lambda)].$$

It remains to add I_1 and I_2 and integrate to obtain

(91) $$\int_{\beta_\sigma} \log \frac{\sigma}{\rho}\, ds = A(\sigma, f') - 2C(\sigma) - A(\sigma, \lambda).$$

16B. To estimate the second term in (83) consider the density ρ with

(92) $$\log \rho(\zeta) = 2 \sum_1^q \log \frac{1}{[\zeta, a_i]} - 2 \log \sum_1^q \log \frac{1}{[\zeta, a_i]} + C,$$

where the second term on the right makes the total mass finite. In fact, if $t = [\zeta, a_i] \to 0$, then $\rho(\zeta) \to \infty$ as rapidly as $t^{-2}(\log t)^{-2}$, and the mass $\int \rho\, d\alpha$ over a t-neighborhood of a_i is dominated by a multiple of $\int_0^t t^{-1}(\log t)^{-2}\, dt$. The constant C is chosen so that the total finite mass is unity. The essence is the first term on the right indicating a heavy concentration of the mass at the q chosen points a_i.

We use the notation

(93) $$B(\sigma) = \sum_1^q B(\sigma, a_i)$$

and have

(94) $$\int_{\beta_\sigma} \log \rho(f(z))\, ds = B(\sigma) - 2 \int_{\beta_\sigma} \log \sum_1^q \log \frac{1}{[f(z), a_i]}\, ds + O(1),$$

where $O(1)$ is independent of σ. The second term on the right dominates

$$-2 \log \sum_1^q \int_{\beta_\sigma} \log \frac{1}{[f(z), a_i]}\, ds = -2 \log B(\sigma) + O(1),$$

and by $B(\sigma) \leq qC(\sigma)$ we obtain

(95) $$\int_{\beta_\sigma} \log \rho\, ds \geq B(\sigma) + O(\log C(\sigma)).$$

17. Exceptional intervals

17A. It remains to estimate the right-hand side of (83). Since δ is the density at a point, we obtain the 4π-fold total mass $M(\sigma)$ distributed on the Riemannian image of Ω_σ by first integrating δ along β_σ and then integrating from 0 to σ:

$$M(\sigma) = 4\pi \int_0^\sigma \int_{\beta_\sigma} \delta \, ds \, d\sigma. \tag{96}$$

On the other hand, this mass is obtained by multiplying the mass element $d\mu(a)$ at a by the number $\nu(\sigma, a) - \nu(0, a)$ of times a is covered and by integrating over the entire S-sphere:

$$M(\sigma) = 4\pi \int_S \nu(\sigma, a) \, d\mu(a) + O(1). \tag{97}$$

We integrate this from 0 to σ, change the order of integration to obtain $A(\sigma, a)$ in the integrand and estimate this upward by $C(\sigma)$. But $C(\sigma)$ does not depend on a, it can be taken in front of the integral sign, and the integral reduces to the total mass which equals one. Thus we obtain a bound for the integral $Q(\sigma)$ of the integral $M(\sigma)$ of the quantity $4\pi \int_{\beta_\sigma} \delta \, ds$:

$$Q(\sigma) = \int_0^\sigma M(\sigma) \, d\sigma < C(\sigma) + O(\sigma). \tag{98}$$

We shall now estimate $M'(\sigma) = 4\pi \int_{\beta_\sigma} \delta \, ds$ separately in the cases $\sigma_\beta = \infty$ and $\sigma_\beta < \infty$.

17B. For $\sigma_\beta = \infty$ and any constant $\alpha \geq 0$ let Δ' be the set of values σ for which $M'(\sigma) \geq e^{\alpha\sigma} M(\sigma)^2$. We choose an arbitrarily small fixed $\sigma_0 > 0$ and let $\sigma > \sigma_0$ in the sequel. Then

$$\int_{\Delta'} e^{\alpha\sigma} \, d\sigma \leq \int_{\Delta'} \frac{dM}{M^2} < \frac{1}{M(\sigma_0)} < \infty.$$

For the set Δ'' of values σ with $M(\sigma) \geq e^{\alpha\sigma} Q(\sigma)^2$ we obtain similarly

$$\int_{\Delta''} e^{\alpha\sigma} \, d\sigma \leq \int_{\Delta''} \frac{dQ}{Q^2} < \frac{1}{Q(\sigma_0)} < \infty.$$

We infer that, for $\sigma \notin \Delta = \Delta' \cup \Delta''$, $M'(\sigma) < e^{3\alpha\sigma} Q(\sigma)^4$ and

$$\log M'(\sigma) < 3\alpha\sigma + 4 \log Q(\sigma). \tag{99}$$

From (98) it follows that for any $\alpha \geq 0$

$$\log \int_{\beta_\sigma} \delta \, ds = O(\sigma + \log C(\sigma)) \tag{100}$$

except perhaps in a set Δ so small that $\int_\Delta e^{\alpha\sigma} \, d\sigma < \infty$.

17C. In the case $\sigma_\beta < \infty$ let

$$\Delta' = \{\sigma \mid M'(\sigma) \geq e^{\alpha/(\sigma_\beta - \sigma)} M(\sigma)^2\}$$

with $\alpha > 0$. Then

$$\int_{\Delta'} e^{\alpha/(\sigma_\beta - \sigma)}\, d\sigma \leq \int_{\Delta'} \frac{dM}{M^2} < \infty.$$

Similarly for

$$\Delta'' = \{\sigma \mid M(\sigma) \geq e^{\alpha/(\sigma_\beta - \sigma)} Q(\sigma)^2\}$$

we have

$$\int_{\Delta''} e^{\alpha/(\sigma_\beta - \sigma)}\, d\sigma \leq \int_{\Delta''} \frac{dQ}{Q^2} < \infty.$$

We conclude for $\sigma \notin \Delta = \Delta' \cup \Delta''$ that

$$\text{(100)}' \qquad \log \int_{\beta_\sigma} \delta\, ds = O\left(\frac{1}{\sigma_\beta - \sigma} + \log C(\sigma)\right).$$

18. Second main theorem

18A. Substitution of (91), (95), and (100) or (100)′ into (83) gives the desired result (Sario [5]):

Theorem. *Let R be a Riemann surface of type R_s. For any finite number $q \geq 3$ of values $a_1, \cdots, a_q$ the sum of the proximity functions $B(\sigma, a_i)$ grows so slowly that, if $\sigma_\beta = \infty$, then*

$$\text{(101)} \qquad \sum_1^q B(\sigma, a_i) < 2C(\sigma) - A(\sigma, f') + A(\sigma, \lambda) + O(\sigma + \log C(\sigma))$$

except perhaps in a set $\Delta = \Delta(\alpha)$ so small that $\int_\Delta e^{\alpha\sigma}\, d\sigma < \infty$ for $\alpha \geq 0$.

If $\sigma_\beta < \infty$, then the term $O(\sigma + \log C(\sigma))$ in (101) is replaced by

$$\text{(101)}' \qquad O\left(\frac{1}{\sigma_\beta - \sigma} + \log C(\sigma)\right)$$

and the resulting inequality holds except in a set Δ with $\int_\Delta e^{\alpha/(\sigma_\beta - \sigma)}\, d\sigma < \infty$ for $\alpha > 0$.

If the metric ds comes, e.g., from the capacity function p_β, then $A(\sigma, \lambda)$ can be replaced by the integrated 4π-fold Euler characteristic of Ω_σ, from which it differs by $O(\sigma)$ (cf. (17) and (47)).

An equivalent formulation of (101) is readily found by substituting for $B(\sigma, a_i)$ from (76). For $\sigma_\beta = \infty$ we have

$$\text{(102)} \qquad (q-2)C(\sigma) < \sum_1^q A(\sigma, a_i) - A(\sigma, f') + A(\sigma, \lambda) + O(\sigma + \log C(\sigma)),$$

whereas for $\sigma_\beta<\infty$ the last term is replaced by (101)′. Both inequalities are valid except perhaps in Δ.

Again, in the case discussed at the end of 14B, inequality (102) can be written in the classical Nevanlinna [22]-Ahlfors [10] form of the second main theorem for meromorphic functions in the plane.

18B. The presence of exceptional intervals Δ in the second main theorem was a consequence of the nature of estimation of $M'(\sigma)$. Since we had to start from an upper bound for the integral of the integral of $M'(\sigma)$, viz. $\int M(\sigma)\,d\sigma<C(\sigma)+O(\sigma)$, a bound cannot always be given for $M'(\sigma)$ for all σ. If, however, $C(\sigma)$ and $A(\sigma,\lambda)$ grow sufficiently slowly, then we shall show that the second main theorem holds without exceptional intervals Δ. We denote $\sum A(\sigma, a_i)$ by $A(\sigma)$.

Theorem. *Suppose $C(\sigma)$ and $A(\sigma,\lambda)$ do not grow more rapidly than $e^{\alpha\sigma}$ for some $\alpha>0$ and $\sigma_\beta=\infty$. Then*

$$(q-2)C(\sigma)+A(\sigma,f') < A(\sigma)+A(\sigma,\lambda)+O(\sigma) \tag{103}$$

holds for every $\sigma>0$.

If $\sigma_\beta<\infty$ and $C(\sigma)$, $A(\sigma,\lambda)$ are dominated by $e^{\alpha/(\sigma_\beta-\sigma)}$ for some $\alpha>0$, then $O(\sigma)$ in (103) is to be replaced by $O(1/(\sigma_\beta-\sigma))$.

Proof. For $\sigma_\beta=\infty$ it follows from $C(\sigma)=O(e^{\alpha\sigma})$ that $\log C(\sigma)=O(\sigma)$, and (103) holds for $\sigma\notin\Delta$. Now let σ be an arbitrary point of an interval in Δ and denote by σ' the right end point of that interval. Then (103) is true for σ'. Since $(q-2)C(\sigma)+A(\sigma,f')$ is an increasing function, we have

$$\begin{aligned}(q-2)C(\sigma)+A(\sigma,f')<A(\sigma)+A(\sigma,\lambda)+[A(\sigma')-A(\sigma)]\\ +[A(\sigma',\lambda)-A(\sigma,\lambda)]+O(\sigma').\end{aligned} \tag{104}$$

From $A(\sigma)=O(e^{\alpha\sigma})$ and the convexity of $A(\sigma)$ it follows that $A'(\sigma)=O(e^{\gamma\sigma})$ and consequently $A(\sigma')-A(\sigma)<c\int_\sigma^{\sigma'}e^{\gamma\sigma}\,d\sigma$ for $\gamma>\alpha$. By the defining property of Δ the integral is $O(1)$. Similarly $A(\sigma',\lambda)-A(\sigma,\lambda)=O(1)$. Furthermore, $\sigma'-\sigma\le\int_\sigma^{\sigma'}e^{\alpha\sigma}\,d\sigma$, hence $\sigma'-\sigma=O(1)$, and we conclude that $O(\sigma')=O(\sigma)$. Statement (103) follows.

If $\sigma_\beta<\infty$, we obtain analogously $\log C(\sigma)=O(1/(\sigma_\beta-\sigma))$ and $A'(\sigma)=O(e^{\gamma/(\sigma_\beta-\sigma)})$ for some $\gamma>\alpha$. The proof, mutatis mutandis, remains valid.

19. Picard points

19A. We know from 14C that $C(\sigma)$ is convex in σ. We now exclude from our consideration the degenerate case by assuming that

$$\lim_{\sigma\to\infty}\frac{C(\sigma)}{\sigma}=\infty \tag{105}$$

if $\sigma_\beta = \infty$. By virtue of (80) this means that we only permit functions with *unbounded spherical area* $S(\sigma)$ of the Riemannian image of R_σ under f. Similarly in the case $\sigma_\beta < \infty$ we make the assumption

$$\limsup_{\sigma \to \sigma_\beta} C(\sigma)(\sigma_\beta - \sigma) = \infty, \tag{105$'$}$$

which implies that $S(\sigma)$ grows more rapidly than $1/(\sigma_\beta - \sigma)$.

An illustrative trivial case is the extended plane punctured at a countable point set. In this region, despite its weak boundary, there trivially are functions with infinitely many Picard values, e.g., the identity function. To exclude such functions of no interest we require that there be, in some sense, an essential singularity on the ideal boundary β of R. The above condition has this effect:

A meromorphic function with property (105) *or* (105)$'$ *comes arbitrarily close to every value a in every boundary neighborhood* $R - R_\sigma$.

To see this suppose $[f(z), a] > \varepsilon$ on $R - R_{\sigma_0}$ for some a, ε and some $\sigma_0 < \sigma_\beta$. Then $B(\sigma, a) < 2 \log \varepsilon^{-1} + O(1)$ and $A(\sigma, a) \le 4\pi\nu(\sigma_0, a)\sigma$ for $\sigma > \sigma_0$. Consequently $C(\sigma) = O(\sigma)$, a contradiction.

We shall return to these questions in IV.6D.

19B. We set

$$\delta(a) = \liminf_{\sigma \to \sigma_\beta} \frac{B(\sigma, a)}{C(\sigma)}, \tag{106}$$

$$\vartheta(a) = \liminf_{\sigma \to \sigma_\beta} \frac{A(\sigma, f_a')}{C(\sigma)}, \tag{107}$$

and

$$\kappa = \liminf_{\sigma \to \sigma_\beta} \frac{A(\sigma, \lambda)}{C(\sigma)}, \tag{108}$$

where, e.g., in the case of the capacity metric, $A(\sigma, \lambda)$ differs by $O(\sigma)$ from $E(\sigma) = 4\pi \int_0^\sigma e(\sigma)\, d\sigma$, with $e(\sigma)$ the Euler characteristic of Ω_σ (see 3A). We have arrived at what we set out to prove,

$$\sum \delta(a) + \sum \vartheta(a) \le 2 + \kappa, \tag{109}$$

which is (67) in the case under consideration.

Having derived this defect and ramification relation using both the general method of §§1 and 2 and the special method of the present section, we again refer to the Introduction, in particular its Summary, where a comparative discussion of the two methods was given.

In the next chapter we shall show that our general method can also be extended to certain mappings into arbitrary Riemann surfaces.

CHAPTER II

MAPPINGS INTO OPEN RIEMANN SURFACES

Our considerations thus far have been limited to the case of closed range surfaces. We now drop this restriction and study given complex analytic mappings of arbitrary (open) Riemann surfaces into arbitrary Riemann surfaces.

In §1 we introduce the necessary tools, the principal functions. We use them in §2 to construct on the range surface S a proximity function $s(\zeta, a)$. The main theorems can then be established in §3 largely in the same manner as for mappings into closed surfaces. There is one significant difference, however. The conformal metric we use has only a finite number of zeros on a closed surface and these zeros could in I.4 be conveniently added to the points $a_1, \cdots, a_q$ chosen on S. In the general case our metric has infinitely many zeros and they bring a new aspect to the theory.

The main results of the present chapter are Theorem 6B on the uniform boundedness from below of $s(\zeta, a)$; Theorem 8B on the area 4π of an arbitrary S in our metric; Theorem 9C, the main theorem for given analytic mappings between arbitrary Riemann surfaces; the affinity relation 10A for mappings into surfaces of finite Euler characteristic; and Theorems 16A and 16B on the vanishing capacity of exceptional sets for arbitrary analytic mappings.

That there actually exist nontrivial mappings into surfaces of arbitrary finite or infinite genus is clear. The projection mapping between two suitable covering surfaces is a simple example.

We shall only touch lightly on the challenging question of the existence of analytic mappings between *given* surfaces. In this direction the field lies wide open for further research. The most significant first steps have been taken by Ozawa [6] and Rodin-Sario [1].

The main method used in this chapter was developed in Sario [10], [13], and [14]. The theorems in 13 to 16 are interesting recent results of Nakai [6].

§1. PRINCIPAL FUNCTIONS

We shall discuss a method of constructing harmonic functions with given singularities and given boundary behavior on an arbitrary Riemann surface. A more detailed description of the problem and its significance will be given in 4B after the necessary concepts have been introduced.

1. Preliminaries

1A. Given a Riemann surface W and a point $\zeta_0 \in W$, consider regions $\Omega \subset \Omega'$ of W containing ζ_0. Let u_Ω be a uniquely determined harmonic function on Ω.

Lemma. *If the Dirichlet integral D_Ω over Ω has the directed limit*

$$\lim_{\Omega \to W} D_\Omega(u_\Omega - u_{\Omega'}) = 0, \tag{1}$$

then $u_\Omega(\zeta) - u_\Omega(\zeta_0)$ converges uniformly in compact subsets to a harmonic limit

$$v(\zeta) = \lim_{\Omega \to W} (u_\Omega(\zeta) - u_\Omega(\zeta_0)). \tag{2}$$

Proof. For $\zeta = \xi_1 + i\xi_2 \in W$ the partial derivative $u_i = u_{\xi_i}$, $i = 1, 2$, of a harmonic u is harmonic. Its value at the center ζ of a parametric disk D of radius ρ is

$$u_i(\zeta) = \frac{1}{\pi\rho^2} \int_D u_i \, dW,$$

where dW is the Euclidean area element of D. On applying the Schwarz inequality and summing on i we obtain

$$|\text{grad } u|^2 \le \frac{1}{\pi\rho^2} \int_D |\text{grad } u|^2 \, dW = \frac{1}{\pi\rho^2} D_D(u).$$

Given a compact set $E \subset W$ cover $E \cup \zeta_0$ with parametric disks $D_m \subset W$, $m = 1, \cdots, N$, such that the disks D'_m concentric with the D_m and of radii $1-\delta$, $0 < \delta < 1$, already cover E. We may assume that $\bigcup_1^N D'_m$ is connected, for if this is not the case, we add a sufficient number of disks. We join ζ_0 to any $\zeta \in E$ by a sequence ζ_j, $j = 1, \cdots, j_\zeta < N$, with $\zeta_{j-1} \cup \zeta_j$ in some D'_m. The line segment d_j from ζ_{j-1} to ζ_j has length < 2, and we find that for u defined on $G = \bigcup D_m$

$$|u(\zeta_j) - u(\zeta_{j-1})| \le 2 \max_{d_j} |\text{grad } u| \le 2\pi^{-1/2}\delta^{-1}(D_G(u))^{1/2}.$$

This implies

$$|u(\zeta) - u(\zeta_0)| \le 2N\pi^{-1/2}\delta^{-1}(D_G(u))^{1/2},$$

and an application to $u(\zeta)=u_\Omega(\zeta)-u_{\Omega'}(\zeta)$ with $G\subset\Omega$ gives the desired result.

1B. The following is a general property of harmonic functions:

Lemma. *Let E be a compact set on an arbitrary Riemann surface W. Consider the class of harmonic functions u on W with*

$$\min_E u \le 0, \qquad \sup_W u \ge 0. \tag{3}$$

There exists a constant q, $0<q<1$, independent of u, such that

$$u \mid E \le q \sup_W u. \tag{4}$$

In the more restricted class of functions u with $\operatorname{sgn} u \mid E \neq \text{const.}$ the lemma implies

$$q \inf_W u \le u \mid E \le q \sup_W u, \tag{5}$$

as can be seen by applying (4) to $-u$. Less sharply, $\max_E |u| \le q \sup_W |u|$.

Proof. If $\sup_W u=0$ or ∞, there is nothing to prove. In other cases we normalize by a multiplicative constant so as to make $\sup_W u=1$. For $v=1-u$ we have $\inf_W v=0$, $\max_E v\ge 1$, hence $v>0$ on W. We are to establish the existence of a $q'\in(0,1)$ such that $\min_E v\ge q'$.

We may again assume that E is connected, and we cover E by a finite number of disks D_m, D'_m with centers z_m as in 1A. By Harnack's inequality $v(\zeta)/v(z_m)$ for $\zeta\in D'_m$ is in the interval (c, c^{-1}), where $c=\delta/(2-\delta)$. For any ζ, ζ' in D'_m we have $v(\zeta)/v(\zeta')\in(c^2, c^{-2})$. There is a point $\zeta_0\in E$ where $v(\zeta_0)\ge 1$ and this point can be connected with any $\zeta\in E$ by a sequence of points $\zeta_1,\cdots,\zeta_{j_\zeta}=\zeta$ with $j_\zeta\le N$, the pair ζ_{j-1}, ζ_j for $j=1,\cdots,j_\zeta$ being in some disk D'_m. Consequently c^{2N} qualifies as q'.

2. Auxiliary functions

2A. By a *bordered Riemann surface* $\overline{W}$ we understand the closure of a subregion W of an (unspecified) Riemann surface W^* such that the relative boundary ∂W of W with respect to W^* consists of a finite number of analytic Jordan curves, and W and W^*-W have the same relative boundary. The set ∂W is referred to as the *border* of $\overline{W}$ (or of W). For an axiomatic definition of these concepts we refer to Ahlfors-Sario [1, p. 117].

Let $\overline{W}$ be a compact bordered Riemann surface with border $\alpha\cup\beta$, where α and β are disjoint sets of analytic Jordan curves. Choose a point

$\zeta_0 \in W$ and a parametric disk D: $r<1$, $r=|\zeta-\zeta_0|$, with $\bar{D}\subset W$. Consider the class G of harmonic functions g in $\overline{W}-\zeta_0$ such that

$$g\,|\,D = \frac{1}{2\pi}\log\frac{1}{r}+e, \tag{6}$$

$$g\,|\,\alpha = 0, \tag{7}$$

$$\int_\alpha dg^* = 1, \tag{8}$$

where e is harmonic in $\bar{D}$.

In the class G we single out the functions g_0, g_1, and g_k defined by the conditions

$$\frac{\partial g_0}{\partial n} = 0 \quad \text{on} \quad \beta, \tag{9}$$

$$g_1\,|\,\beta = c \text{ (const.)}, \tag{10}$$

$$g_k = hg_0+kg_1, \tag{11}$$

h, k being real constants with $h+k=1$.

2B. We shall first establish an extremal property of g_1 by evaluating the Dirichlet integral $D(g-g_1)$ over W. We denote by a, a_0, a_1 the values $e(\zeta_0)$ for g, g_0, g_1, respectively, and set $B(g)=\int_\beta g\,dg^*$, $B(g,\hat{g})=\int_\beta g\,d\hat{g}^*$ for $g, \hat{g}\in G$.

Lemma. *The function g_1 minimizes $B(g)+a$ in G.*

More generally, g_k minimizes $B(g)+(k-h)a$, the value of the minimum is $k^2a_1-h^2a_0$, and the deviation from this minimum is $D(g-g_k)$:

$$B(g)+(k-h)a = k^2a_1-h^2a_0+D(g-g_k). \tag{12}$$

Proof. We start with

$$D(g-g_k) = B(g)+B(g_k)-B(g,g_k)-B(g_k,g).$$

To evaluate

$$B(g_k) = hk[B(g_0,g_1)-B(g_1,g_0)]$$

we take a circle γ: $|\zeta-\zeta_0|=r_0<1$ and write $C(g,\hat{g})=\int_\gamma g\,d\hat{g}^*$ for $g, \hat{g}\in G$. Then

$$B(g_k) = hk[C(g_0,g_1)-C(g_1,g_0)].$$

Here we have oriented β and γ so as to leave W and $|\zeta-\zeta_0|<r_0$ to the left. On denoting $(1/2\pi)\log(1/r)$ by ρ we obtain

$$\begin{aligned}B(g_k) &= hk[C(\rho+e_0,\rho+e_1)-C(\rho+e_1,\rho+e_0)]\\ &= hk[C(e_0,\rho)-C(e_1,\rho)]\\ &= hk[e_1(\zeta_0)-e_0(\zeta_0)] = hk(a_1-a_0),\end{aligned}$$

where e_i signifies e for g_i with $i=0, 1$.

In the same manner we derive the equations

$$\begin{aligned}B(g,g_k) &= k(a_1-a),\\ B(g_k,g) &= h(a-a_0),\end{aligned}$$

and conclude that (12) holds.

2C. Now let $\overline{W}$ be a noncompact bordered Riemann surface with compact border α, and denote by Ω a regular subregion with border $\alpha\cup\beta_\Omega$. Here α and β_Ω are again disjoint sets of analytic Jordan curves.

Let G be the class of harmonic functions g on $\overline{W}-\zeta_0$ defined by conditions (6) to (8). The functions $g_{0\Omega}$, $g_{1\Omega}$, and $g_{k\Omega}$ in $\bar\Omega-\zeta_0$ are defined by obvious modifications of conditions (9) to (11). We shall show that these functions converge uniformly in compact subsets to unique harmonic functions g_0, g_1, g_k on $\overline{W}-\zeta_0$ as $\Omega\to W$.

2D. Let $\bar\Omega\subset\Omega'$ and apply (12) to $g=g_{0\Omega'}$ and $g_k=g_{0\Omega}$. Since $B_\Omega(g_{i\Omega'})\le B_{\Omega'}(g_{i\Omega'})=0$ for $i=0, 1$, we obtain

$$D_\Omega(g_{0\Omega'}-g_{0\Omega}) \le a_{0\Omega}-a_{0\Omega'}.$$

Similarly $g=g_{1\Omega'}$, $g_k=g_{1\Omega}$ gives

$$D_\Omega(g_{1\Omega'}-g_{1\Omega}) \le a_{1\Omega'}-a_{1\Omega},$$

and $g=g_{\Omega 0}$, $g_k=g_{1\Omega}$ leads to

$$D_\Omega(g_{0\Omega}-g_{1\Omega}) \le a_{0\Omega}-a_{1\Omega}.$$

We conclude that the directed limits

$$a_i = \lim_{\Omega\to W} a_{i\Omega}, \tag{13}$$

$i=0, 1$, exist. Consequently

$$\lim_{\Omega\to W} D_\Omega(g_{i\Omega'}-g_{i\Omega}) = 0. \tag{14}$$

We reflect $\overline{W}$, $\bar\Omega$, $\bar\Omega'$ across α to form the doubles $\hat W$, $\hat\Omega$, $\hat\Omega'$ and infer that for $\bar\Omega\subset\Omega'$ the corresponding integral $D_{\hat\Omega}$ of the harmonic extension $\hat g_{i\Omega'}-\hat g_{i\Omega}$ to $\hat\Omega$ of $g_{i\Omega'}-g_{i\Omega}$ also tends to zero. By virtue of the triangle inequality, $D_{\hat\Omega_0}(\hat g_{i\Omega}-\hat g_{i\Omega'})\to 0$ for any fixed Ω_0 as Ω, Ω' exhaust W

independently of each other. Since $\hat{g}_{i\Omega}=0$ on α, one infers by Lemma 1A that $\hat{g}_{i\Omega}$ converges uniformly on compact subsets of $\hat{W}-\zeta_0-\zeta_0^*$ (ζ_0^* is the reflection of ζ_0), hence on those of $W-\zeta_0$. We obtain the limiting functions g_0, g_1, and

$$g_k = \lim_{\Omega\to W} g_{k\Omega} = hg_0+kg_1. \tag{15}$$

One can show that (12) remains valid on the noncompact $\overline{W}$ but we shall not need this information.

3. Linear operators

3A. Let $\overline{W}$ again be a compact bordered Riemann surface with border $\alpha\cup\beta$. Let f be a harmonic function on α and consider the class H of harmonic functions u on $\overline{W}$ with

$$u \,|\, \alpha = f, \tag{16}$$

$$\int_\alpha du^* = 0. \tag{17}$$

In H let the functions u_0, u_1, u_k be defined by

$$\frac{\partial u_0}{\partial n} = 0 \quad \text{on} \quad \beta, \tag{18}$$

$$u_1 \,|\, \beta = c \text{ (const.)}, \tag{19}$$

$$u_k = hu_0+ku_1, \tag{20}$$

where $h+k=1$. Let $A(u)=\int_\alpha u\, du^*$, $B(u)=\int_\beta u\, du^*$. Then the Dirichlet integral is $D(u)=B(u)-A(u)$.

Lemma. *The functions u_0 and u_1 minimize $D(u)$ and $A(u)+B(u)$ in H, respectively.*

More generally,

$$B(u)+(k-h)A(u) = k^2A(u_1)-h^2A(u_0)+D(u-u_k). \tag{21}$$

The proof is the same as that of Lemma 2B, if g, g_k, a, a_k are replaced by u, u_k, $A(u)$, $A(u_k)$.

3B. If $\overline{W}$ is noncompact with compact border α, the family H consists, by definition, of harmonic functions u on $\overline{W}$ with $u=f$ on α and $\int_\alpha du^*=0$. In the same manner as in 2D one proves that $A(u_{0\Omega})$ decreases, $A(u_{1\Omega})$ increases with increasing Ω, and that the directed limits

$$A(u_i) = \lim_{\Omega\to W} A(u_{i\Omega}),$$

$i=0, 1$, exist. The integrals $D_\Omega(u_{i\Omega'}-u_{i\Omega})$ tend to zero and we have the limiting functions u_0, u_1, and

(22) $$u_k = \lim_{\Omega\to W} u_{k\Omega} = hu_0+ku_1.$$

Lemma 3A can be shown to remain valid for the limiting functions, but this property will not be needed.

3C. The operator L_k is defined by

(23) $$u_k = L_k f.$$

It satisfies the conditions

(24) $$L_k f\,|\,\alpha = f,$$

(25) $$L_k(c_1 f_1+c_2 f_2) = c_1 L_k f_1 + c_2 L_k f_2,$$

(26) $$\int_\alpha d(L_k f)^* = 0,$$

(27) $$\min f \le L_k f \le \max f$$

for compact $\overline{W}$. By virtue of uniform convergence the same is true for noncompact $\overline{W}$.

The operator L_k has an integral representation in terms of g_k. If $\overline{W}$ is compact, we apply Green's formula to u_i, g_i along $\beta-\alpha-\gamma$, and let γ shrink to ζ_0. The resulting equations $u_i(\zeta_0)=\int_\alpha f\,dg_i^*$ give

(28) $$u_k(\zeta_0) = \int_\alpha f\,dg_k^*.$$

Because of the uniform convergence of both $u_{k\Omega}$ and $g_{k\Omega}$ we have:

Lemma. *On an arbitrary bordered Riemann surface with compact border* α

(29) $$L_k f(\zeta_0) = \int_\alpha f\,dg_k^*,$$

where α *is oriented so as to leave* W *to its right.*

3D. On an arbitrary open Riemann surface W let $\overline{W}_0$ be a compact bordered subregion with border α_0. Let $\overline{W}_1$ with border $\alpha_1\subset W_0$ be the complement of a regular subregion of W. For a real-valued function $f\in C$ on α_0 let Lf be the solution of the Dirichlet problem in $\overline{W}_0$. The operator L_k acting on $Lf\,|\,\alpha_1$ gives on $\overline{W}_1$ the harmonic function Kf with

(30) $$K = L_k L.$$

The nth iterate of K is denoted by K^n.

Let q be the constant of Lemma 1B applied to the compact set α_1 on the Riemann surface W_0.

Lemma. *If* sgn $K^i f \mid \alpha_1 \neq$ const., $i=1, \cdots, n$, *then*

$$q^n \min f \le K^n f \mid \alpha_0 \le q^n \max f. \tag{31}$$

For $n=1$ this follows from (5) and (27). For each iteration we obtain another factor q.

4. An integral equation

4A. Orient α_0 and α_1 so that they leave $W_0 \cap W_1$ to the left and right, respectively. In $\overline{W}_0 \cap \overline{W}_1$ let ω be the harmonic function with conjugate $\omega^* = s$ such that $\omega \mid \alpha_1 = 0$, $\omega \mid \alpha_0 = \text{const.} > 0$, $\int_{\alpha_1} ds = 1$. Choose branches of s with the property that the curves

$$\alpha_0 : \zeta = t_0(s), \qquad \alpha_1 : \zeta = t_1(s) \tag{32}$$

are traced as s increases from 0 to 1. Let $\partial/\partial n$ stand for the normal derivative on α_0 and α_1 interior to $W_0 \cap W_1$.

Denote by $g(\zeta, \zeta_0)$ the Green's function on $\overline{W}_0$ with the singularity at $\zeta_0 \in W_0$. Given harmonic functions f_0, f_1 on α_0, α_1, respectively, the L-operators have the following integral representations:

$$Lf_0(t_1(s)) = \int_0^1 f_0(t_0(y)) \frac{\partial g(t_0(y), t_1(s))}{\partial n} dy, \tag{33}$$

$$L_k f_1(t_0(x)) = \int_0^1 f_1(t_1(s)) \frac{\partial g_k(t_1(s), t_0(x))}{\partial n} ds. \tag{34}$$

We introduce the kernel

$$K(x, y) = \int_0^1 \frac{\partial g(t_0(y), t_1(s))}{\partial n} \frac{\partial g_k(t_1(s), t_0(x))}{\partial n} ds \tag{35}$$

and have

$$Kf_0(t_0(x)) = \int_0^1 K(x, y) f_0(t_0(y)) \, dy. \tag{36}$$

4B. Given a harmonic function σ on $\overline{W}_1$ and an operator L_k we wish to construct on W a harmonic function p such that $p - \sigma = L_k(p - \sigma)$ on $\overline{W}_1$. The function will be called the *principal* function corresponding to σ, L_k. Here W can be an open Riemann surface W^* punctured at a finite number of points ζ_j, and W_1 can consist of neighborhoods D_j of the ζ_j and a neighborhood D_β of the ideal boundary β of W^*. The harmonic function σ in D_j may have an isolated singularity at ζ_j, and on D_β it may be a function behaving arbitrarily as one approaches β. Thus we are dealing with the problem of constructing a harmonic function on W^* with given singularities and a prescribed behavior near the ideal boundary.

We may assume that $\sigma \mid \alpha_1 = 0$, for otherwise we can replace σ by $\sigma - L_k\sigma$. It is necessary that the flux $\int_{\alpha_1} d\sigma^*$ vanish, for so do $\int_{\alpha_1} dp^*$ and $\int_{\alpha_1} d(L_k(p-\sigma))^*$.

Our problem will be solved if we can find a p on α_0 such that $p - Kp = \sigma$ on α_0, for then the function defined as Lp in W_0 and $\sigma + Kp$ on W_1 is the required one. That is, on α_0 we have the equation

$$p = \sigma + Kp \tag{37}$$

or, more precisely,

$$p(t_0(x)) = \sigma(t_0(x)) + \int_0^1 K(x, y)p(t_0(y))\, dy. \tag{38}$$

Thus we are dealing with a Fredholm integral equation. It is known that its solution is $p = \sum_0^\infty K^n\sigma$ provided the series converges uniformly. In fact, the K-operator can then be applied term by term and gives $Kp = \sum_1^\infty K^n\sigma = p - \sigma$.

4C. For the convergence proof we first observe that, by virtue of Green's formula,

$$\int_{\alpha_0} u\, ds = \int_{\alpha_1} u\, ds \tag{39}$$

for any harmonic function u on $\overline{W}_0 \cap \overline{W}_1$ with $\int_{\alpha_1} du^* = 0$. The functions σ, Lf_0, L_kf_1, and Kf_0 qualify as u.

It is easy to see that

$$\int_{\alpha_1} K^i\sigma\, ds = 0 \tag{40}$$

for all $i \geq 0$. In fact, for $i = 0$ this is so by our assumption $\sigma \mid \alpha_1 = 0$. Suppose then (40) holds for $i = m-1$. Then $\int_{\alpha_0} K^{m-1}\sigma\, ds = 0$. Here the integrand can be replaced by L acting on it, whence $\int_{\alpha_1} LK^{m-1}\sigma\, ds = 0$. The operator L_k can now be applied to the integrand, and (40) follows for $i = m$.

We conclude that $K^i\sigma$, $i = 1, 2, \cdots$, is not of constant sign on α_1. Lemma 3D gives to $p \mid \alpha_0 = \sum_0^\infty K^n\sigma \mid \alpha_0$ the upper bound $\max_{\alpha_0} \sigma/(1-q)$ and the lower bound $\min_{\alpha_0} \sigma/(1-q)$. By the maximum-minimum principle the same bounds hold for $p \mid W_0$, hence for $p \mid \alpha_1$ and $p - \sigma \mid \alpha_1$ and a fortiori for $p - \sigma$ on all of W_1.

We have established the following result (Sario [13]):

Theorem. *For a harmonic function* σ *in* $\overline{W}_1$ *with* $\sigma \mid \alpha_1 = 0$, $\int_{\alpha_1} d\sigma^* = 0$, *the solution* $p \mid \alpha_0 = \sum_0^\infty K^n \sigma$ *of the Fredholm integral equation* (38) *gives the principal function* p *on* W *such that* $p - \sigma = L_k p$ *on* W_1 *and*

$$\frac{\min_{\alpha_0} \sigma}{1-q} \le p \mid W_0 \le \frac{\max_{\alpha_0} \sigma}{1-q}, \tag{41}$$

$$\frac{\min_{\alpha_0} \sigma}{1-q} \le p - \sigma \mid W_1 \le \frac{\max_{\alpha_0} \sigma}{1-q}. \tag{42}$$

These bounds only depend on $\min_{\alpha_0} \sigma$ and $\max_{\alpha_0} \sigma$, not on σ otherwise. This makes it possible to give bounds simultaneously for uniformly bounded families of functions σ. In essence, if $\sigma \mid \alpha_0$ is $O(1)$, then so are $p \mid W_0$ and $p - \sigma \mid W_1$, all uniformly. It is here that the significance of principal functions in value distribution theory lies, as we shall presently see.

§2. PROXIMITY FUNCTIONS ON ARBITRARY RIEMANN SURFACES

We are now able to construct on an arbitrary Riemann surface S a proximity function, i.e., a function that describes the nearness of a generic point ζ to a given point a. This function is required to have two properties: it must tend logarithmically to infinity as ζ tends to a, and it must remain positive or at least uniformly bounded from below as ζ and a vary on the surface. In the classical case of the plane region such functions are immediately available, e.g., $\log^+ |\zeta - a|^{-1}$ and $\log [\zeta, a]^{-1}$. In contrast, on an abstract Riemann surface the construction of a proximity function turns into an essential part of the theory.

The basic idea of our construction is the same as in I.1. On an arbitrary Riemann surface S, open or closed, take two points ζ_0 and ζ_1. Let t_0 be a harmonic function with positive and negative singularities at ζ_0, ζ_1, respectively, and with $t_0 = L_1 t_0$ in a neighborhood of the ideal boundary β of S. In a sense, t_0 then has a constant value on β. The function $s_0 = \log(1 + e^{t_0})$ continues to have a positive logarithmic pole at ζ_0 but it is bounded from below on S. To s_0 we add the harmonic function $t = t(\zeta, a)$ with positive and negative singularities at a and ζ_0, respectively, and with $t = L_1 t$ near β. The singularities at ζ_0 then cancel, and the sum $s(\zeta, a) = s_0(\zeta) + t(\zeta, a)$ has as singularity only the positive logarithmic pole at a. The symmetry $s(\zeta, a) = s(a, \zeta)$, essential for our purposes, will be achieved by normalizing t at ζ_0 so that $t(\zeta, a) - 2 \log |\zeta - \zeta_0| \to s_0(a)$ as $\zeta \to \zeta_0$. The problem is to show that $s(\zeta, a) \ge O(1)$ uniformly for all ζ and a.

5. Boundedness of auxiliary functions

5A. Let S be an open Riemann surface and consider a bordered neighborhood $\bar{D}_\beta$ of the ideal boundary β of S, with compact border ∂D_β. Let u, v be harmonic functions in $\bar{D}_\beta$ with $u = L_1 u$, $v = L_1 v$. Then

$$\int_{\partial D_\beta} u\, dv^* - v\, du^* = 0. \tag{43}$$

In fact, we have shown that any function $u = L_1 u$ is the uniform limit of harmonic functions u_Ω defined on compact bordered subregions $\bar{\Omega} \subset \bar{D}_\beta$ with borders $\beta_\Omega - \partial D_\beta$. Furthermore $u_\Omega = \text{const.}$ on β_Ω with vanishing flux, and consequently

$$\int_{\partial D_\beta} u_\Omega\, dv_\Omega^* - v_\Omega\, du_\Omega^* = \int_{\beta_\Omega} u_\Omega\, dv_\Omega^* - v_\Omega\, du_\Omega^* = 0.$$

Statement (43) follows from uniform convergence.

5B. Let S be an arbitrary Riemann surface, open or closed. Throughout our presentation we denote by D a parametric disk and by D' and D'' increasingly smaller disks concentric with D. On $\bar{S}'' = S - D''$ let g_a be the auxiliary g_1-function constructed in 2 with a positive logarithmic pole at $a \in S''$, vanishing on the curve $\partial D''$ and with boundary behavior L_1. However, for the coefficient of the logarithmic singularity we now take 2 instead of $1/2\pi$. The function is symmetric:

Lemma. *For $a, b \in S''$,*

$$g_a(b) = g_b(a). \tag{44}$$

Proof. Let D_a and D_b be parametric disks about a and b with $g_a = -2 \log r + O(1)$ in D_a, and $g_b = -2 \log r + O(1)$ in D_b. Here and later r stands for the distance of the generic point ζ from the center of the parametric disk in question. Take level lines γ_a and γ_b of g_a and g_b about a and b in D_a and D_b, respectively. Let $\bar{D}_\beta$ be a bordered boundary neighborhood of S that does not meet $\bar{D}'' \cup \bar{D}_a \cup \bar{D}_b$. By Green's formula we first have

$$\int_{\partial D'' + \partial D_\beta - \gamma_a - \gamma_b} g_a\, dg_b^* - g_b\, dg_a^* = 0. \tag{45}$$

By using (43), the vanishing of g_a and g_b on $\partial D''$, and the harmonicity of g_a and g_b in D_b and D_a, respectively, we infer that

$$\int_{\gamma_a} g_b\, dg_a^* = \int_{\gamma_b} g_a\, dg_b^*.$$

The mean value theorem gives

$$4\pi g_b(a) = 4\pi g_a(b).$$

5C. The following property of g_a will be instrumental in our reasoning:
Lemma. *Let E be a compact set in S'' and $O \subset S''$ an open set containing E. Then*

(46) $$g_a \mid E = O(1) \quad \text{uniformly for} \quad a \in S'' - O.$$

In the proof we make use of the symmetry property $g_a(t) = g_t(a)$. We must show that

$$g_t \mid S'' - O = O(1) \quad \text{uniformly for } t \in E.$$

Cover E with a finite set of disks K_i in O such that slightly smaller concentric disks K_i' already cover E. Decompose E into compact subsets E_i contained in K_i'. It suffices to show the uniform boundedness for $t \in E_i$. In applying Theorem 4C here and below we choose L_1 for L_k in each component of W_1. For W we take here $S'' - t$ and let W_1 consist of three components, $K_i - t$, $D - D''$, and a bordered neighborhood D_β, with compact ∂D_β, of the ideal boundary β of S. For W_0 choose $S' - K_i' - D_\beta'$, where $S' = S - D'$ and D_β' is a bordered neighborhood of β with compact $\partial D_\beta' \subset D_\beta$. In $K_i - t$ take $\sigma = 2 \log (|1 - \zeta \bar{t}|/|\zeta - t|)$, and in $D - D''$, $\sigma = 2 \log r$. In D_β we set $\sigma = 0$. The conditions $\sigma \mid \partial W_1 = 0$ and $\int_{\partial W_1} d\sigma^* = 0$ are obviously satisfied, $\sigma \mid \partial W_0 = O(1)$ uniformly, and by Theorem 4C, $p - \sigma$ and $p \mid W_0$ are $O(1)$ uniformly. We normalize on $\partial D''$. Since p on it is c, a constant, we take $g_t = p - c$ and have $g_t \mid S'' - O = O(1)$. In fact, this is true on $(D - D'') \cup D_\beta$, and the rest of the set $S'' - O$ is a subset of W_0.

In the proof we took the liberty of including or excluding contours in our regions to simplify notations. We also tacitly assumed that $\bar{D}$ does not meet $\bar{K}_i$. If this is not the case, then we replace D, D' by smaller concentric disks $C \supset C'$ containing $\bar{D}''$ and with $\bar{C} \cap \bar{K}_i = \varnothing$.

6. Uniform boundedness from below of $s(\zeta, a)$

6A. We are now ready to carry out the construction of the proximity function according to our program in the introduction to §2. To form t_0 take $W = S - \zeta_0 - \zeta_1$ and $W_1 = (D_0 - \zeta_0) \cup (D_1 - \zeta_1)$, where the parametric disks D_0, D_1 are centered at ζ_0, ζ_1. In $D_0 - \zeta_0$ choose $\sigma = -2 \log r$, in $D_1 - \zeta_1$, $\sigma = 2 \log r$. We tacitly have also a neighborhood D_β of the ideal boundary β, but since we can choose $\sigma = 0$ there we no longer write it down here or in later applications of Theorem 4C. For W_0 take $S - D_0' - D_1'$, the meaning of the primes being as in 5B. Since $\sigma \mid \partial W_0 = O(1)$ and the flux vanishes, we infer that $p - \sigma = O(1)$ and $p \mid W_0 = O(1)$. The principal function p is taken as t_0. Then $t_0 \mid D_0 = -2\log r + O(1)$ and it follows that $s_0 \mid D_0 = \log(1 + r^{-2}e^{O(1)}) = -2\log r + \log(r^2 + O(1))$.

We have established the following estimate:

Lemma. *$s_0 \mid D_0 = -2\log r + O(1)$ and $s_0 \geq 0$ on S.*

6B. The construction of $s = s_0 + t$ will depend on the location of a. Let D and $\tilde{D}$ be disks disjoint from each other and from D_0''. Consider three cases: (I) $a \in D_0''$, (II) $a \in S - D_0'' - D$, (III) $a \in S - D_0'' - \tilde{D}$. The union of the three sets is S, and it suffices to establish a uniform lower bound for s separately in each of the three cases. The third case can be dispensed with since it is the same as the second.

Case I. $a \in D_0''$. Take $W = S - \zeta_0 - a$, $W_1 = D_0 - \zeta_0 - a$, and $W_0 = S - D_0'$. Set $\sigma = 2\log(r|1 - \zeta\bar{a}|/|\zeta - a|)$. Then $\sigma \mid \partial W_0$ is $O(1)$ and so are $p - \sigma$ and $p \mid W_0$. The normalization is at ζ_0 where $p - 2\log r$ tends to the limit $-2\log|a| + c(a)$ as $r \to 0$, with $c(a) = O(1)$ uniformly for $a \in D_0''$. By Lemma 6A this limit is $s_0(a) + c_1(a)$, where again $c_1(a) = O(1)$. The function $t = p - c_1$ has the required normalization $t(\zeta, a) - 2\log|\zeta - \zeta_0| \to s_0(a)$ as $\zeta \to \zeta_0$ (this will entail the symmetry of $s = s_0 + t$ in 7D). Moreover, since $|1 - \zeta\bar{a}|/|\zeta - a| > 1$, we have $t \mid D_0 > 2\log r + O(1)$ and $t \mid S - D_0 = O(1)$. On combining this with Lemma 6A we obtain $s \mid D_0 > O(1)$, $s \mid S - D_0 > O(1)$, hence $s > O(1)$ uniformly for $a \in D_0''$.

Case II. $a \in S - D_0'' - D$. On $S'' = S - D''$ we have $-g_{\zeta_0} \mid D_0 = 2\log r + O(1)$, $-g_{\zeta_0} \mid S'' - D_0 = O(1)$. On applying Lemma 5C to $E = \partial D' \cup \zeta_0$, $O = (D - D'') \cup D_0''$ we obtain $g_a \mid S'' \geq 0$, $g_a \mid \partial D' = O(1)$, $g_a \mid \zeta_0 = O(1)$. Consequently the restriction of $g_a - g_{\zeta_0}$ to D_0 is $> 2\log r + O(1)$; to $S'' - D_0$, $> O(1)$; to $\partial D'$, $O(1)$; and at ζ_0 we have $g_a - g_{\zeta_0} - 2\log r \to c(a) = O(1)$ as $\zeta \to \zeta_0$, uniformly in a.

As the last application of Theorem 4C we take $W = S - \zeta_0 - a$, $W_1 = S'' - \zeta_0 - a$, $W_0 = D'$, and $\sigma = g_a - g_{\zeta_0}$ in W_1. Then $\sigma \mid \partial W_0 = O(1)$. The normalization is at ζ_0 where $p - 2\log r \to c_1(a) = O(1)$. Take $t = p + s_0(a) - c_1(a)$. Since $s_0(a) - c_1(a) > O(1)$, we conclude that $t \mid D_0 > 2\log r + O(1)$, $t \mid S'' - D_0 > O(1)$, and $t \mid D'' = O(1)$. Adding t to s_0 gives $s \mid D_0 > O(1)$, $s \mid S - D_0 > O(1)$.

We have established the following result (Sario [10]):

Theorem. *The proximity function $s(\zeta, a)$ is uniformly bounded from below for all ζ, a on an arbitrary Riemann surface S.*

7. Symmetry of $s(\zeta, a)$

7A. We shall show that for $a, b \in S$, $s(a, b) = s(b, a)$. We could use the same reasoning as in I.1B. However, we also wish to prove that $s(\zeta, a)$ is the uniform limit of corresponding functions constructed on exhausting subregions. This convergence is needed to show that the total area (see 8B) in the metric we shall use is precisely 4π even for an open S.

Suppose first that $\bar{S}$ is a bordered Riemann surface with compact border β and let D_a, D_0 be parametric disks about a, ζ_0, with disjoint closures in S. Let V be the class of harmonic functions v on $S-\zeta_0-a$ with the same singularities and normalization as t:

$$v \mid \bar{D}_a - a = -2 \log r + h,$$
$$v \mid \bar{D}_0 - \zeta_0 = 2 \log r + k,$$

where h, k are harmonic in $\bar{D}_a$, $\bar{D}_0$, and $k(\zeta_0) = s_0(a)$. Set $c = 4\pi h(a)$. In V single out the functions v_0, v_1 determined by

$$\frac{\partial v_0}{\partial n} \mid \beta = 0, \qquad v_1 \mid \beta = \text{const.},$$

and set for real λ

(47) $$v_\lambda = (1-\lambda)v_0 + \lambda v_1.$$

The quantities h, k, c corresponding to v_λ will be denoted by h_λ, k_λ, c_λ. For $v, v' \in V$ we write $B(v) = \int_\beta v\, dv^*$, $B(v, v') = \int_\beta v\, dv'^*$.

Lemma. *The function v_λ minimizes $B(v)+(2\lambda-1)c$ in V:*

(48) $$B(v)+(2\lambda-1)c = \lambda^2 c_1 - (1-\lambda)^2 c_0 + D(v-v_\lambda).$$

Proof. The Dirichlet integral of $v-v_\lambda$ over S is $D(v-v_\lambda) = B(v)+B(v_\lambda) - B(v, v_\lambda) - B(v_\lambda, v)$. Let $A_i(v)$, $i=1, 2$, be the integral $\int v\, dv^*$ along ∂D_a, ∂D_0, respectively, and similarly for $A_i(v, v')$. In the same manner as in 2B we obtain

$$B(v_\lambda) = (1-\lambda)\lambda \sum_1^2 (A_i(v_0, v_1) - A_i(v_1, v_0)).$$

Here the first summand is $A_1(h_0, -2 \log r) - A_1(h_1, -2 \log r) = c_1 - c_0$. Because of the normalization at ζ_0 the second summand vanishes and we have

$$B(v_\lambda) = (1-\lambda)\lambda(c_1 - c_0).$$

Similarly

$$B(v, v_\lambda) = \lambda B(v, v_1) = \lambda(c_1 - c),$$
$$B(v_\lambda, v) = (1-\lambda)B(v_0, v) = (1-\lambda)(c-c_0).$$

Equality (48) follows.

7B. Now let S be an arbitrary open Riemann surface, and Ω a regular subregion with border β_Ω. Let $v_{\lambda\Omega}$ be the function (47) constructed on Ω as above.

Lemma. *The directed limit*

$$v_\lambda = \lim_{\Omega \to S} v_{\lambda\Omega} \tag{49}$$

exists and the convergence is uniform in compact subsets of $S-a-\zeta_0$.

Proof. The train of thought is, in essence, the same as in 2D. Let $\bar{\Omega} \subset \Omega'$ and indicate by primes the quantities corresponding to Ω' and $\beta_{\Omega'}$. We apply (48) to $v=v_0'$, $v_\lambda=v_0$ and obtain

$$B(v_0')-c_0' = -c_0+D(v_0'-v_0). \tag{50}$$

Analogously

$$B(v_1')+c_1' = c_1+D(v_1'-v_1) \tag{51}$$

and

$$B(v_1)-c_1 = -c_0+D(v_1-v_0). \tag{52}$$

From these equations and from the relations $B(v_i') \le B'(v_i')=0$, $B(v_1)=0$, $D \ge 0$, we infer that c_1 increases while c_0 decreases with increasing Ω, and $c_1 \le c_0$ for every Ω. A fortiori the limit $c_i = \lim c_{i\Omega}$ exists, with the obvious meaning of $c_{i\Omega}$. This implies

$$\lim_{\Omega \to S} D_\Omega(v_{i\Omega'}-v_{i\Omega}) = 0. \tag{53}$$

In view of the normalization $v_{i\Omega'}(\zeta_0)-v_{i\Omega}(\zeta_0)=0$ the asserted convergence follows from Lemma 1A.

For later reference (8B) we let $\Omega' \to S$ while keeping Ω fixed in (51), and obtain

$$B_\Omega(v_1)+c_1 = c_{1\Omega}+D_\Omega(v_1-v_{1\Omega}).$$

For $\Omega \to S$ this gives

$$\lim_{\Omega \to S} B_\Omega(v_1) = 0. \tag{54}$$

7C. Although both v_0 and v_1 were needed in the convergence proof, we shall only make use of v_1 in the sequel.

Lemma. *The function t constructed in* 6B *and the limiting function* v_1 *in* (49) *are identical.*

Proof. By definition, $t=L_1 t$, and by virtue of the uniqueness of principal functions it suffices to show that $v_1 = L_1 v_1$. Let

$$W_1 = (D_a-a) \cup (D_0-\zeta_0) \cup D_\beta$$

and let Ω contain $S-W_1$. Denote by $L_{1\Omega}$ the L_1-operator acting on functions on $\alpha_1 = \partial W_1$ and providing us with harmonic functions on

$\bar{\Omega} \cap \overline{W}_1$. Then $v_{1\Omega} = L_{1\Omega} v_{1\Omega}$ and we are to prove that $\lim L_{1\Omega} v_{1\Omega} = L_1 v_1$. On α_1 we have $v_{1\Omega} \to v_1$ and consequently

$$L_{1\Omega} v_{1\Omega} - L_{1\Omega} v_1 = L_{1\Omega}(v_{1\Omega} - v_1) \to 0,$$

i.e., $\lim L_{1\Omega} v_{1\Omega} = \lim L_{1\Omega} v_1$. By the definition of L_1 the latter expression is indeed $L_1 v_1$.

The above reasoning for t can also be applied to show that

$$t_0 = \lim_{\Omega \to S} t_{0\Omega}, \tag{55}$$

where $t_{0\Omega}$ is the t_0-function constructed on Ω.

7D. It is now easy to see that $s = s_0 + t$ is symmetric (Sario [14]):

Lemma. *For any* $a, b \neq \zeta_0$

$$s(a, b) = s(b, a). \tag{56}$$

Proof. Again first suppose $\bar{S}$ is bordered compact with border β. Let $\alpha_a, \alpha_b, \alpha_0$ be the peripheries of parametric disks about a, b, ζ_0 and set $t_a = t(\zeta, a)$, $t_b = t(\zeta, b)$ with $t_a \mid \beta = \text{const.}$, $t_b \mid \beta = \text{const.}$ Then

$$\int_{\beta - \alpha_0 - \alpha_a - \alpha_b} t_a\, dt_b^* - t_b\, dt_a^* = 0.$$

Here $\int_\beta = 0$ and, in the same fashion as in 7A, we obtain $\int_{\alpha_0} = 4\pi(s_0(a) - s_0(b))$. Analogous computations give $\int_{\alpha_a} = 4\pi t_b(a)$ and $\int_{\alpha_b} = -4\pi t_a(b)$. We infer that

$$s_0(a) + t(a, b) = s_0(b) + t(b, a).$$

This is (56).

If S is noncompact the statement follows from the above and the uniform convergence of the approximating functions formed on the $\Omega \subset S$.

8. Conformal metric

8A. As in I.1C we shall now form a conformal metric in terms of t_0 and show that even for open surfaces S the total area is 4π.

Let the area element be $d\omega = \lambda^2 dS$, where

$$\lambda^2 = \Delta s = \Delta s_0 = \frac{e^{t_0} |\text{grad}\, t_0|^2}{(1 + e^{t_0})^2}. \tag{57}$$

For $x \in (-\infty, \infty)$ denote by $\alpha(x, t_0)$ the level line $t_0 = x$ on $S - \zeta_0 - \zeta_1$. When x is near $-\infty$ or ∞, then $\alpha(x, t_0)$ is compact and encircles ζ_1 or ζ_0, respectively. On an open S, α is noncompact for some values x.

Given a regular region $\Omega \subset S$ containing ζ_0, ζ_1 set $\alpha_\Omega(x, t_0) = \alpha(x, t_0) \cap \Omega$.

Lemma. *On a noncompact S we have for $x \in (-\infty, \infty)$*

$$\lim_{\Omega \to S} \int_{\alpha_\Omega(x, t_0)} dt_0^* \leq 4\pi \quad \text{a.e.} \tag{58}$$

Proof. If the statement is false, then there exists a constant $\varepsilon > 0$, a regular region Ω_0, and a value x_0 with $\alpha_0 = \alpha_{\Omega_0}(x_0, t_0)$, such that $\int_{\alpha_0} dt_0^* > 4\pi + 3\varepsilon$, grad $t_0 \mid \alpha_0 \neq 0$, and α_0 is not tangent to $\partial\Omega_0$. Moreover, there is a $\delta > 0$ with grad $t_0 \mid G \neq 0$ and $\int_\gamma |dt_0^*| = \langle\varepsilon\rangle$, where

$$G = \Omega_0 \cap \{\zeta \mid |t_0(\zeta) - x_0| < \delta\},$$

$\gamma = \bar{G} \cap \partial\Omega_0$ consists of disjoint closed arcs γ_1, γ_2 joined by α_0, and $\langle\varepsilon\rangle$ stands for a quantity in the interval $(-\varepsilon, \varepsilon)$.

Because of the uniform convergence $t_{0\Omega} \to t_0$ there exists a regular region $\Omega \supset \Omega_0$ such that

$$\int_{\alpha_0} dt_{0\Omega}^* = \int_{\alpha_0} dt_0^* + \langle\varepsilon\rangle,$$

$$\int_\gamma |dt_{0\Omega}^*| = \int_\gamma |dt_0^*| + \langle\varepsilon\rangle = \langle 2\varepsilon\rangle,$$

and

$$|t_{0\Omega} - t_0| \mid G < \delta.$$

For any arc $\alpha \subset G$ from γ_1 to γ_2 we have

$$\int_\alpha dt_{0\Omega}^* = \int_{\alpha_0} dt_{0\Omega}^* + \langle 2\varepsilon\rangle = \int_{\alpha_0} dt_0^* + \langle 3\varepsilon\rangle > 4\pi.$$

On the other hand, the arc $\alpha = \alpha_{\Omega_0}(x_0, t_{0\Omega})$ gives

$$\int_\alpha dt_{0\Omega}^* \leq \int_{\alpha_\Omega(x_0, t_{0\Omega})} dt_{0\Omega}^* = 4\pi.$$

This contradiction proves the lemma.

8B. After these preparations we are ready to show (Sario [14]):

Theorem. *The total area of S is*

$$\int_S d\omega = 4\pi. \tag{59}$$

Proof. Let $\alpha_1 \colon t_0 = x_1$ and $\alpha_2 \colon t_0 = x_2$ be level lines such that the sets $\zeta_1 \cup \{\zeta \mid t_0(\zeta) \leq x_1\}$ and $\zeta_0 \cup \{\zeta \mid t_0(\zeta) \geq x_2\}$ are compact. We know from (54) that $\int_{\partial\Omega} t_0 dt_0^* \to 0$ as $\Omega \to S$. For this reason the Dirichlet integral of t_0 over the region with $x_1 < t_0 < x_2$ is

$$D(t_0) = \lim_{\Omega \to S} \int_{\alpha_2 - \alpha_1 + \partial\Omega} t_0 dt_0^* = 4\pi(x_2 - x_1)$$

$$= \lim_{\Omega \to S} \int_{x_1}^{x_2} \int_{\alpha_\Omega(x,\, t_0)} dt_0^* \, dx.$$

From this and from Lemma 8A we conclude that

$$\lim_{\Omega \to S} \int_{\alpha_\Omega(x,\, t_0)} dt_0^* = 4\pi \quad \text{a.e.}$$

The theorem follows on integrating (57).

8C. We finally remark that λ *is finite everywhere and its only zeros are those of* grad t_0. In fact, the only points that need checking are ζ_0 and ζ_1. At ζ_0 we have

$$\lambda^2 = e^{-t_0} \frac{|\text{grad } t_0|^2}{(e^{-t_0}+1)^2}.$$

As $r = |\zeta - \zeta_0| \to 0$, $t_0 \sim 2 \log (r^{-1})$, $|\text{grad } t_0|^2 \sim 4r^{-2}$, $e^{-t_0} \sim r^2$, and consequently $\lambda^2 \sim 4$, which shows that λ has neither an infinity nor a zero at ζ_0. In the vicinity of ζ_1 we similarly have for $r = |\zeta - \zeta_1| \to 0$

$$\lambda^2(\zeta) \sim \frac{r^2 \cdot (4/r^2)}{(1+r^2)^2} \sim 4.$$

The Gaussian curvature K corresponding to our metric is, as in the case of a closed S, constantly 1. As a by-product we thus have (Sario [10]):

Theorem. *There exists a metric (with zeros) of finite total mass and constant Gaussian curvature on an arbitrary Riemann surface.*

§3. ANALYTIC MAPPINGS

With the uniform boundedness from below of the proximity function established and the total area of the related metric evaluated, the scene is set for the main theorems on given analytic mappings between arbitrary Riemann surfaces.

For nondegenerate mappings the main theorems imply the affinity relation, while for arbitrary mappings the set on which the proximity function is unbounded is shown to be of vanishing capacity. The existence of mappings between given Riemann surfaces is briefly discussed.

9. Main theorems

9A. Let R be an arbitrary open Riemann surface and R_0 a parametric disk with border β_0. Take an adjacent regular region Ω with border $\beta_0 \cup \beta_\Omega$, $\bar{R}_0 \cap \bar{\Omega} = \beta_0$, and a harmonic function u on $\bar{\Omega}$ such that $u \mid \beta_0 = 0$, $u \mid \beta_\Omega = k$ (const.) with $\int_{\beta_0} du^* = 1$. Let β_h be the level line $u^{-1}(h)$, $h \in [0, k]$, and denote by Ω_h the region $u^{-1}((0, h))$.

Suppose $\zeta = f(z)$ is a complex analytic mapping of R into a closed or open Riemann surface S. For $a \in S$ let $\nu(h, a)$ be the number of inverse images z_j of a in $R_h = \bar{R}_0 \cup \Omega_h$, counted with their multiplicities. To study value distribution under f we use the same fundamental A-, B-, and C-functions as for a closed S:

$$A(h, a) = 4\pi \int_0^h \nu(h, a)\, dh,$$

$$B(h, a) = \int_{\beta_h - \beta_0} s(f(z), a)\, du^*,$$

$$C(h) = \int_0^h \int_{R_h} d\omega(f(z))\, dh.$$

An application of Stokes' formula to the functions $s(f(z), a)$ and $h - u(z)$ over Ω_h less disks about the z_j shrinking to their centers gives in the same manner as in I.2:

Theorem. *The equality*

$$A(k, a) + B(k, a) = C(k) \tag{60}$$

is a universal property of analytic mappings of arbitrary Riemann surfaces into arbitrary Riemann surfaces.

Thus the $(A+B)$-affinity is again the same for all points a. Moreover, this invariant sum equals $C(k)$, the integral of the area of the Riemannian image of R_h over S.

As in I.4A we set

$$\varphi_i(h) = \int_0^h \varphi_{i-1}(x)\, dx,$$

$i = 1, 2$, with $\varphi_0 = \varphi$, for any integrable function φ in $[0, k]$. In contrast with I.4A, however, $\psi(h)$ shall now signify

$$\psi(h) = \sum_1^q \psi(h, a_i).$$

The points ζ_0, ζ_1 are taken different from the given $a_1, \cdots, a_q$, and R_0 is then chosen so that $f(\beta_0)$ does not meet any of these points. Now we cannot include the zeros of λ in the sum as they may be infinite in number. Accordingly, we have

$$A_2(k)+B_2(k) = qC_2(k). \tag{61}$$

To study defect points we are to estimate B_2.

9B. To this end we now set

$$\sigma(\zeta) = \exp\left[\sum_1^q s(\zeta, a_i)-2\log\left(\sum_1^q s(\zeta, a_i)+\text{const.}\right)\right] \tag{62}$$

and introduce the mass distribution $dm=\sigma\, d\omega$ on S. The total mass $m=\int_S dm$ is finite. To see this we delete from S a neighborhood of each a_i. The remaining region clearly has finite mass, since σ has a finite maximum on it and the mass element dm is majorized by a multiple of $d\omega=\lambda^2 dS$ which gives a finite area to S. In the deleted neighborhoods the finiteness of the mass is seen in the same manner as in I.4B.

9C. The density on S of our mass distribution is $\sigma\lambda^2$ which in the $(u+iu^*)$-plane induces the density $\sigma\mu^2$ with

$$\mu(z) = \lambda(f(z))|f'(z)|\,|\text{grad}\, u(z)|^{-1}. \tag{63}$$

In terms of μ we again set

$$\begin{aligned} F(h) &= \int_{\beta_h} \log(\sigma\mu^2)\,du^*, \\ G(h) &= -\int_{\beta_h} \log\mu^2\,du^*, \end{aligned} \tag{64}$$

and have

$$B_2(h) < F_2(h)+G_2(h)+2[\log(C(h)+O(1))]_2+O(h^2). \tag{65}$$

Exactly the same reasoning as in 5A and 5B gives the following estimates, where $e(h)$ is the Euler characteristic of Ω_h, and $E(h)=4\pi\int_0^h e(h)\,dh$:

$$F_2(h) < h^2(\log C(h)+O(\log h)) \tag{66}$$

and

$$G_2(h) = -A_2(h,\lambda)-A_2(h,f')+E_2(h)+2C_2(h)+O(h^3). \tag{67}$$

It is in these estimates that the crucial properties of $s(\zeta, a)$ are needed: symmetry and uniform boundedness from below.

On combining these estimates we obtain the following main theorem (Sario [10]):

Theorem. *Given an analytic mapping f of an arbitrary open Riemann surface R into an arbitrary open or closed Riemann surface S, the inequality*

$$(68) \quad (q-2)C_2(k) < \sum_1^q A_2(k, a_i) - A_2(k, f') - A_2(k, \lambda) + E_2(k) + O(k^3 + k^2 \log C(k))$$

is valid for every regular subregion Ω of R.

10. Affinity relation

10A. In the present no. we restrict our attention to certain special mappings. General mappings will be considered in 12 to 16.

We suppose that

$$(69) \qquad \lim_{R_k \to R} \frac{k^3 + k^2 \log C(k)}{C_2(k)} = 0$$

and introduce the defect of a,

$$\alpha(a) = 1 - \limsup \frac{A_2(k, a)}{C_2(k)}.$$

Here, as in the sequel, the limit is a directed limit as $R_k \to R$. Rao's condition I.(31) is again seen to be sufficient for (69).

The ramification index is defined by

$$\beta(a) = \liminf \frac{A_2(k, f_a')}{C_2(k)},$$

where the numerator counts the orders of branch points above a, and the total ramification is

$$\sum \beta(a) \leq \liminf \frac{A_2(k, f')}{C_2(k)},$$

with the sum extended over all points a of S.

For the zeros of λ we set

$$\gamma(a) = \liminf \frac{A_2(k, \lambda_a)}{C_2(k)},$$

where the numerator counts the points of $f(R_k)$ covering a zero of λ at a. Only a finite number of zeros are covered by $f(R_k)$ and we can consider the sum

$$\sum \gamma(a) \leq \liminf \frac{A_2(k, \lambda)}{C_2(k)}$$

extended over all a on S.

Finally we have the Euler index

$$\eta = \liminf \frac{E_2(k)}{C_2(k)}.$$

As a consequence of (68) we obtain (Sario [10]):

Affinity relation. *For mappings with* (69) *between Riemann surfaces*

$$\sum \alpha(a) + \sum \beta(a) + \sum \gamma(a) \leq 2 + \eta. \tag{70}$$

All terms are obviously positive, and for $\eta < \infty$ we obtain estimates for each of the three terms. First, the defect sum and consequently the number of Picard points is dominated by $2+\eta$. The same bound is valid for the total ramification; in particular, there can be at most $4+2\eta$ totally ramified points, i.e., points covered exclusively by branch points. Finally, the sum of the γ-indices is bounded by $2+\eta$.

10B. As a special case we obtain results corresponding to Chapter I for a closed range surface S. Since grad t_0 forms a vector field on $S-\zeta_0-\zeta_1$, the number of its zeros, i.e., zeros of λ, is the Euler characteristic e_S+2. We now add to our arbitrarily chosen points $a_1, \cdots, a_q$ these $2g$ zeros, g the genus, replace q in (68) by $q+e_S+2$, and arrive at

$$\sum \alpha + \sum \beta \leq \eta - e_S.$$

In particular, the number of Picard points is at most the excess of the Euler index η over the Euler characteristic of the range surface S. For meromorphic functions on an arbitrary Riemann surface or for Gaussian mappings of arbitrary minimal surfaces the bound for the defect sum is $2+\eta$. In the most special case of meromorphic functions in the plane, $\eta=0$ and we are back to the classical bound 2 for the defect sum and the total ramification.

11. Existence of mappings

11A. As stated in the beginning of 10A we are here dealing with special cases even if S is open. In fact, the affinity relation (70) is, in general, meaningful only if S has finite Euler characteristic. To see this we shall prove (Rodin-Sario [1]):

Theorem. *Suppose the number* $\nu(\lambda)$ *of the zeros of metric* (57) *on* S *is infinite. Then there do not exist mappings satisfying* (69) *and with* $\eta < \infty$.

Proof. To the points $a_1, \cdots, a_q$ of 9A we add the zeros $\{a_{q+k}\}_{k=1}^{\infty}$ of our metric. For a positive integer l we now take the mass distribution $dm = \sigma\, d\omega$ with

$$\sigma(\zeta) = \exp\Big[\sum_1^{q+l} s(\zeta, a_i) - 2\log\Big(\sum_1^{q+l} s(\zeta, a_i) + \text{const.}\Big)\Big].$$

As in 9B and 9C we obtain for any nonconstant analytic $f\colon R \to S$

$$(q+l-2)C_2(k) < \sum_1^{q+l} A_2(k, a_i) - A_2(k, f') - A_2(k, \lambda) + E_2(k) + O(k^3 + k^2 \log C(k)). \tag{71}$$

Consider the function $\tilde{A}_2(k, \lambda)$ defined by

$$\sum_1^{q+l} A_2(k, a_i) - A_2(k, \lambda) = \sum_1^{q} A_2(k, a_i) - \tilde{A}_2(k, \lambda).$$

It counts the number of times f takes on an a_j with $j \geq q+l+1$ in R_k and is therefore nonnegative. By virtue of (69) and (71)

$$\sum_1^q \left(1 - \frac{A_2(k, a_i)}{C_2(k)}\right) + \frac{A_2(k, f')}{C_2(k)} + \frac{\tilde{A}_2(k, \lambda)}{C_2(k)} + l - 2 < \frac{E_2(k)}{C_2(k)} + o(k).$$

We suppress two positive terms, take inferior limits on both sides and obtain

$$\sum_1^q \alpha(a_i) + l - 2 \leq \eta. \tag{72}$$

Here l can be arbitrarily large and we conclude that $\eta = \infty$. This proves the theorem.

The important value distribution problem for the case $\eta = \infty$ is open for further research. Some results on this difficult question have been recently obtained by Ozawa [1].

11B. The natural question now arises: When do there exist any nonconstant analytic mappings from a given R into a given S?

The problem can be formulated as follows. Let O_S be the class of Riemann surfaces R such that every analytic mapping of R into S reduces to a constant, i.e., to a mapping into a single point of S. Then the problem is to determine the class O_S.

Recently Ozawa in [4] to [8] obtained interesting results of which we state here a criterion for a surface R to belong to O_S. Let $M(R)$ be the class of nonconstant meromorphic functions on R and denote by $P(f)$ the number of Picard values of $f \in M(R)$. We associate with R the quantity

$$P(R) = \sup_{M(R)} P(f). \tag{73}$$

Clearly $P(R) \geq 2$. In fact, according to the Behnke-Stein theorem [1] there exists a nonconstant analytic function f on every R, and $P(e^f) \geq 2$.

Theorem. *If $P(R) < P(S)$, then $R \in O_S$.*

Proof. Suppose there existed a nonconstant analytic mapping g of R into S. For any $f \in M(S)$, $f \circ g \in M(R)$ and we would have

$$P(f) \leq P(f \circ g) \leq P(R).$$

It would follow that $P(S) \leq P(R)$, a contradiction.

11C. Some further observations are equally immediate (cf. Ahlfors-Sario [1, Ch. IV]):

(a) In all cases $O_S \subset O_{AB}$.

(b) If S is the Riemann sphere, then $O_S = \varnothing$.

(c) If S is the finite plane, then O_S consists of all closed surfaces.

(d) If $S \notin O_G$, O_{HP}, O_{HB}, or O_{AB}, then $O_S \supset O_G$, O_{HP}, O_{HB}, or O_{AB}, respectively. In particular, $S \notin O_{AB}$ implies $O_S = O_{AB}$.

Thus the problem seems to have a close connection with the classification theory of Riemann surfaces. However, the problem of characterizing O_S in the general case is not within our reach. It also is outside the scope of value distribution theory proper, which starts from a *given* analytic mapping.

12. Area of exceptional sets

12A. We proceed to arbitrary given mappings. The conclusions in 10 were valid only for mappings satisfying (69) and meaningful only for those with $\eta < \infty$. What can be said about exceptional points if no restrictions are imposed on the mappings? Then the Picard-nature is not a good characterization of the exceptional behavior, since even an open set of the range surface S might be left uncovered. A more suitable criterion is the B-function. As a consequence of Theorem 6B we shall show in 12 to 16 that even in the most general case (an appropriate modification of) the B-function can be unbounded, in a sense, only on a small set S. In the present no. we consider the area of the exceptional set and in 13 to 16, its capacity.

First we observe:

Lemma. *For arbitrary analytic mappings f*

$$\int_S A(k, a)\, d\omega(a) = 4\pi C(k). \tag{74}$$

We know that $\int_S \nu(k, a)\, d\omega(a)$ is the area of the Riemannian image of R_k under f, and $C(k)$ is the integral of that area. The statement follows on interchanging the order of integration.

12B. We consider the function

$$B_h(a) = \int_{\beta_h} s(f(z), a)\, du^*, \tag{75}$$

which differs from the B-function by a constant:

$$B(h, a) = B_h(a) - B_0(a). \tag{76}$$

The following result (Sario [14]) holds for all mappings f, whether conditions (69) and $\eta < \infty$ are satisfied or not (cf. App. I.22):

Theorem. *For every analytic mapping f of an arbitrary Riemann surface R into another arbitrary Riemann surface S the set of exceptional points is so small that*

$$\int_S B_h(a)\, d\omega(a) = O(1), \tag{77}$$

where the bound for $O(1)$ is independent of $\Omega \subset R$.

Proof. Clearly the integral on the left-hand side of (77) is independent of h, and we are to show that

$$\int_S B_0(a)\, d\omega(a) = O(1) \tag{78}$$

for all Ω.

Cover $f(\beta_0)$ by parametric disks D_i, $i=2, \cdots, N$, such that $\bigcup \bar{D}_i$ is disjoint from the parametric disks D_0, D_1 about ζ_0, ζ_1, and slightly smaller disks D_i' concentric with the D_i already cover $f(\beta_0)$. Decompose β_0 into closed arcs β_i, with $f(\beta_i) \subset D_i'$, $\bigcup \beta_i = \beta_0$.

We know that

$$s(\zeta, a) = s(a, \zeta) = s_0(a) + t(a, \zeta).$$

For $\zeta \in f(\beta_i)$ the principal function $t(a, \zeta)$ is constructed from the singularities

$$\sigma \,|\, D_i = 2 \log \left| \frac{1 - a\bar{\zeta}}{a - \zeta} \right|, \qquad \sigma \,|\, D_0 = 2 \log |a - \zeta_0|.$$

On $\partial D_i'$ and on the periphery $\partial D_0'$ of a disk D_0', slightly smaller than D_0, $\sigma = O(1)$. For this reason $t \,|\, D_i = \sigma + O(1)$, $t \,|\, D_0 = \sigma + O(1)$, $t \,|\, S - D_i - D_0 = O(1)$, and consequently $\int_S t(a, \zeta)\, d\omega(a) = O(1)$.

Similarly $t_0 \,|\, D_0 = -2 \log |a - \zeta_0| + O(1)$, $t_0 \,|\, D_1 = 2 \log |a - \zeta_1| + O(1)$, $t_0 \,|\, S - D_0 - D_1 = O(1)$, and we infer for $s_0 = \log(1 + e^{t_0})$ that $\int_S s_0(a)\, d\omega(a) = O(1)$. To summarize, $\int_S s(a, \zeta)\, d\omega(a) = O(1)$ and

$$\int_S B_0(a)\, d\omega(a) = \int_{\beta_0} \int_S s(a, \zeta)\, d\omega(a)\, du^* = O(1).$$

12C. An immediate consequence is (cf. App.I.22):

Corollary. *The set E of points of S for which* lim inf $B_k(a)$ *is infinite has vanishing area* $\int_E d\omega$.

Here the integral is defined as the infimum of integrals over integrable sets containing E. An interesting question is: To what extent can this corollary be sharpened? We shall show in 13 to 16 that E has zero capacity. The proof is due to Nakai [6].

13. Decomposition of $s(\zeta, a)$ in subregions

13A. To evaluate the capacity of the set of exceptional points it will be necessary to develop potential theory with respect to the kernel $s(\zeta, a)$. To this end we shall first show that $s(\zeta, a)$ is continuous as a function of two variables. We start by decomposing $s(\zeta, a)$ in regular subregions.

We recall that $s(\zeta, a) = s_0(\zeta) + t(\zeta, a)$. By definition, a function is finitely continuous if it is a continuous map into the real line $(-\infty, \infty)$. The function $t(\zeta, a)$ has the following properties, according as we consider it as a function of ζ or a or (ζ, a):

(α) $\zeta \to t(\zeta, a)$ is harmonic on $S - a - \zeta_0$,

(β) $a \to t(\zeta, a)$ is finitely continuous on $S - \zeta$,

(γ) $(\zeta, a) \to t(\zeta, a)$ is bounded from below on $U \times V$.

Here U and V are parametric disks on S chosen as follows: given an arbitrary point (ζ', a') of $S \times S$ with $\zeta' \neq a'$ and $\zeta' \neq \zeta_0$, the disks U, V are centered at ζ', a', respectively, such that $\bar{U} \cap \bar{V} = \varnothing$ and $\zeta_0 \notin \bar{U}$.

Property (α) follows from the construction of $t(\zeta, a)$, and (β) is a consequence of the symmetry of $s(\zeta, a)$ giving $t(\zeta, a) = -s_0(\zeta) + s(a, \zeta)$. To see ($\gamma$) we note that s is, as we showed in 6B, bounded from below on $S \times S$, and therefore by the above equality $t(\zeta, a)$ is bounded from below on $U \times V$.

From (α), (β), (γ) and Harnack's inequality we deduce that $(\zeta, a) \to t(\zeta, a)$ is finitely continuous on $U \times V$, and in particular at (ζ', a').

The function $(\zeta, a) \to s(\zeta, a)$ is finitely continuous at $(\zeta', a') \in S \times S$ with $\zeta' \neq a'$.

In fact, by the symmetry $s(\zeta, a) = s(a, \zeta)$ we may assume that $\zeta' \neq \zeta_0$. Since $(\zeta, a) \to t(\zeta, a)$ is finitely continuous at (ζ', a'), and $\zeta \to s_0(\zeta)$ is finitely continuous at ζ', $(\zeta, a) \to s(\zeta, a) = s_0(\zeta) + t(\zeta, a)$ has the asserted property.

13B. Let Ω be a regular region in S and let g_Ω be the Green's function of Ω. Take an arbitrary but then fixed point $(\zeta, a) \in \Omega \times \Omega$ and set for $b \in \Omega$

$$u(b, a) = s(b, a) - 2g_\Omega(b, a).$$

Then $\Delta_b u(b, a) = \Delta_b s(b, a) - 2\Delta_b g_\Omega(b, a) = \Delta_b s_0(b) = \lambda^2(b)$ on $\Omega - a$. The singularities of s and $2g_\Omega$ cancel and we have $\Delta_b u(b, a) = \lambda^2(b)$ on all of Ω.

Let U_ε be a disk with center ζ and radius ε such that $\bar{U}_\varepsilon \subset \Omega$. By Green's formula

$$\int_{\beta_\Omega - \alpha_\varepsilon} u(b, a)\, dg_\Omega^*(b, \zeta) - \int_{\beta_\Omega - \alpha_\varepsilon} g_\Omega(b, \zeta)\, du^*(b, a)$$
$$= \int_{\Omega - \bar{U}_\varepsilon} (u(b, a)\Delta_b g_\Omega(b, \zeta) - g_\Omega(b, \zeta)\Delta_b u(b, a))\, dS_b,$$

where β_Ω and α_ε are borders of Ω and U_ε, respectively, and dS_b is the area element of S. We obtain

$$\int_{\beta_\Omega} u(b, a)\, dg_\Omega^*(b, \zeta) - \int_{\alpha_\varepsilon} u(b, a)\, dg_\Omega^*(b, \zeta) + \int_{\alpha_\varepsilon} g_\Omega(b, \zeta)\, du^*(b, a)$$
$$= -\int_{\Omega - \bar{U}_\varepsilon} \lambda^2(b) g_\Omega(b, \zeta)\, dS_b$$

and, on letting $\varepsilon \to 0$,

$$-u(\zeta, a) = \frac{1}{2\pi}\int_{\beta_\Omega} u(b, a)\, dg_\Omega^*(b, \zeta) + \frac{1}{2\pi}\int_\Omega \lambda^2(b) g_\Omega(b, \zeta)\, dS_b.$$

For short we set

$$G_\Omega(\zeta) = \frac{1}{2\pi}\int_\Omega \lambda^2(b) g_\Omega(b, \zeta)\, dS_b,$$

$$H_\Omega(\zeta, a) = \frac{1}{2\pi}\int_{\beta_\Omega} u(b, a)\, dg_\Omega^*(b, \zeta),$$

$$-v_\Omega(\zeta, a) = G_\Omega(\zeta) + H_\Omega(\zeta, a).$$

We have proved:

Lemma. *The decomposition*

$$s(\zeta, a) = 2g_\Omega(\zeta, a) + v_\Omega(\zeta, a)$$

is valid on $\Omega \times \Omega$.

To establish the joint continuity of $s(\zeta, a)$ we shall next show that $v_\Omega(\zeta, a)$ is finitely continuous on $\Omega \times \Omega$. This is achieved by demonstrating that G_Ω and H_Ω possess this property.

14. Joint continuity of $s(\zeta, a)$

14A. Let ζ' be a fixed point in Ω and let U be a disk with radius 1 and center ζ' such that $\bar{U} \subset \Omega$. Denote by U_r the disk $|b - \zeta'| < r$ in U with $0 < r < 1$, and by g the Green's function on U. Then

$$g(b, \zeta) = \log\left|\frac{1 - \bar{\zeta}b}{b - \zeta}\right| \le \log\frac{1}{|b - \zeta|} + \log 2.$$

Since $g_\Omega(b, \zeta) - g(b, \zeta)$ is finitely continuous on $U \times U$,

$$M = \sup |g_\Omega(b, \zeta) - g(b, \zeta)| < \infty,$$

where the supremum is taken for $(b, \zeta) \in U_{1/2} \times U_{1/2}$. For $0 < \varepsilon < 1/6$ and $\zeta \in U_\varepsilon$ the disks U_ε and $|b - \zeta| < 2\varepsilon$ are contained in $U_{1/2}$. Therefore

$$g_\Omega(b, \zeta) \le \log \frac{1}{|b-\zeta|} + M', \qquad M' = M + \log 2$$

for $|b - \zeta| < 2\varepsilon$ and $\zeta \in U_\varepsilon$.

Let $m = \sup_{b \in U} \lambda^2(b) < \infty$. Then

$$\int_{|b-\zeta|<2\varepsilon} \lambda^2(b) g_\Omega(b, \zeta)\, dS_b \le m \int_0^{2\pi} \int_0^{2\varepsilon} \left(\log \frac{1}{r} + M'\right) r\, dr\, d\theta.$$

For this reason

$$\int_{|b-\zeta|<2\varepsilon} \lambda^2(b) g_\Omega(b, \zeta)\, dS_b = O(\varepsilon)$$

for $\zeta \in U_\varepsilon$. Since the union Ω of $\Omega - U_\varepsilon$ and the disk $|b - \zeta'| < 2\varepsilon$ is identical with the union of $\Omega - U_\varepsilon$ and the disk $|b - \zeta''| < 2\varepsilon$ for $\zeta'' \in U_\varepsilon$, we obtain

$$|G_\Omega(\zeta') - G_\Omega(\zeta'')| \le \int_{\Omega - \bar{U}_\varepsilon} \lambda^2(b) |g_\Omega(b, \zeta') - g_\Omega(b, \zeta'')|\, dS_b$$
$$+ \sum_{\zeta = \zeta', \zeta''} \int_{|b-\zeta|<2\varepsilon} \lambda^2(b) g_\Omega(b, \zeta)\, dS_b$$
$$= \int_{\Omega - \bar{U}_\varepsilon} \lambda^2(b) |g_\Omega(b, \zeta') - g_\Omega(b, \zeta'')|\, dS_b + O(\varepsilon)$$

for $\zeta'' \in U_\varepsilon$. Clearly $g_\Omega(b, \zeta') - g_\Omega(b, \zeta'') \to 0$ uniformly on $\Omega - U_\varepsilon$ as $\zeta'' \to \zeta'$ and therefore

$$\limsup_{\zeta'' \to \zeta'} |G_\Omega(\zeta') - G_\Omega(\zeta'')| < O(\varepsilon).$$

On letting $\varepsilon \to 0$ we conclude that

$$\lim_{\zeta \to \zeta'} G_\Omega(\zeta) = G_\Omega(\zeta').$$

14B. We have seen in 13A that $(b, a) \to u(b, a) = s(b, a) - 2g_\Omega(b, a) = s(b, a)$ is finitely continuous on $\beta_\Omega \times \Omega$ and the same is true of the coefficients of $dg^*_\Omega(b, \zeta)$ as functions of (b, ζ) for some fixed local parameters on β_Ω. Thus $H_\Omega(\zeta, a)$ is finitely continuous on $\Omega \times \Omega$. We have reached the following result (Nakai [6]):

Theorem. *The function $s(\zeta, a)$ is continuous on $S \times S$. Specifically, for every regular region Ω of S there exists a symmetric finitely continuous function $v_\Omega(\zeta, a)$ on $\Omega \times \Omega$ such that*

$$s(\zeta, a) = 2g_\Omega(\zeta, a) + v_\Omega(\zeta, a). \tag{79}$$

15. Consequences

15A. Let μ be a regular Borel measure with compact support S_μ in S. The s-potential s^μ of the measure μ is defined by

$$s^\mu(\zeta) = \int s(\zeta, a)\, d\mu(a).$$

Since $c = \inf_{(\zeta, a) \in S \times S} s(\zeta, a) - 1 > -\infty$,

$$\tilde{s}(\zeta, a) = s(\zeta, a) - c$$

is strictly positive on $S \times S$. It is more convenient to use $\tilde{s}$ than s as the kernel of potentials. Accordingly we consider the potential

$$\tilde{s}^\mu(\zeta) = \int \tilde{s}(\zeta, a)\, d\mu(a),$$

which is strictly positive if $\mu \not\equiv 0$, finitely continuous and subharmonic in $S - S_\mu$, and lower semicontinuous on S.

Let Ω be a regular region containing S_μ and let g_Ω be the Green's function of Ω. As the first application of Theorem 14B we have

$$\tilde{s}^\mu(\zeta) = 2g_\Omega^\mu(\zeta) + v_\Omega(\zeta), \tag{80}$$

where g_Ω^μ is the Green's potential

$$g_\Omega^\mu(\zeta) = \int g_\Omega(\zeta, a)\, d\mu(a)$$

on Ω, and $v_\Omega(\zeta) = \int (v_\Omega(\zeta, a) - c)\, d\mu(a)$ is a finitely continuous function on Ω.

An immediate consequence of (80) is that the potentials $\tilde{s}^\mu$ satisfy the *continuity principle*:

If $\tilde{s}^\mu \mid S_\mu$ is finitely continuous on S_μ, then $\tilde{s}^\mu$ is finitely continuous on S.

More precisely, for $\zeta' \in S_\mu$ we have as $\zeta \to \zeta'$

$$\limsup_{\zeta \in S - S_\mu} \tilde{s}^\mu(\zeta) \leq \limsup_{\zeta \in S_\mu} \tilde{s}^\mu(\zeta). \tag{81}$$

We also deduce from (80):

A set of $\tilde{s}$-capacity ($= s$-capacity) zero is identical with a set of g_Ω-capacity zero and hence of logarithmic capacity zero (App. I.7).

15B. The most important consequence of (80) is the following, known as the *fundamental existence theorem* in potential theory. For the proof see, e.g., Ninomiya [1], Kishi [1], or Nakai [4].

Lemma. *Let K be a compact subset of S and let $u(\zeta)$ be a strictly positive finite upper semicontinuous function on K. There exists a measure μ with compact support S_μ in K such that*

$$\tilde{s}^\mu(\zeta) \geq u(\zeta)$$

on K except for a set of capacity zero and with the further property

$$\tilde{s}^\mu(\zeta) \leq u(\zeta)$$

everywhere on S_μ.

In particular $\tilde{s}^\mu(\zeta) = u(\zeta)$ on S_μ except for a set of capacity zero.

15C. We turn to the maximum principle for our $\tilde{s}$-potentials. The behavior of $\tilde{s}^\mu$ near S_μ is regulated by (81), but the behavior at the ideal boundary β of S is unknown. We can only state that for any compact set K

$$M(K) = \sup_{a \in K} \limsup_{\zeta \to \beta} \tilde{s}(\zeta, a) < \infty. \tag{82}$$

In fact, we have $\tilde{s}(\zeta, a) = s_0(\zeta) + t(\zeta, a) - c$, where $s_0(\zeta)$ is bounded on S outside of a neighborhood of ζ_0 and $(\zeta, a) \to t(\zeta, a)$ is continuous. Moreover, $\zeta \to t(\zeta, a)$ is harmonic on $S - a - \zeta_0$ and has L_1-behavior at the ideal boundary of S.

From this, inequality (81), and the maximum principle for subharmonic functions we deduce the following form of *the maximum principle for $\tilde{s}^\mu$*:

Lemma. *If $\tilde{s}^\mu \leq M$ on S_μ, then*

$$\tilde{s}^\mu \leq \max\{M, \mu(S_\mu) \cdot M(S_\mu)\} \quad on \quad S.$$

If $S \in O_G$, then the ideal boundary is negligible and we have the usual maximum principle:

Corollary. *If $S \in O_G$ and $\tilde{s}^\mu \leq M$ on S_μ, then $\tilde{s}^\mu \leq M$ on S.*

16. Capacity of exceptional sets

16A. Let f be an arbitrary analytic mapping of R into S. As in 12B we set (cf. App.I.22):

$$\tilde{B}_\kappa(a) = \int_{\beta_\kappa} \tilde{s}(f(z), a)\, du^*, \tag{83}$$

where β_k is the boundary of $R_k = \bar{R}_0 \cup \Omega$. Then $\tilde{B}_k(a)$ is essentially the same as $B_h(a)$ of 12B with $h = k$:

(84) $$B_k(a) = \tilde{B}_k(a) + c, \qquad c = \inf_{S \times S} s(\zeta, a) - 1.$$

We set

$$E = \{a \mid a \in S, \liminf_{R_k \to R} \tilde{B}_k(a) = \infty\}.$$

Assume E has positive capacity. By Lemmas 15B and 15C we can find a nonzero measure μ such that $S_\mu \subset E$ and

$$\tilde{s}^\mu(\zeta) \leq \max\{1, \mu(S_\mu) \cdot M(S_\mu)\} \equiv m$$

on S. Then by Fubini's theorem

$$\int \tilde{B}_k(a)\, d\mu(a) = \int_{S_\mu} \left\{ \int_{\beta_k} \tilde{s}(f(z), a)\, du^* \right\} d\mu(a) = \int_{\beta_k} \tilde{s}^\mu(f(z))\, du^* \leq m.$$

By Fatou's lemma (cf. App.I.21)

$$\infty = \int (\liminf_{R_k \to R} \tilde{B}_k(a))\, d\mu(a) \leq \liminf_{R_k \to R} \int \tilde{B}_k(a)\, d\mu(a) \leq m.$$

This contradiction shows that E must be of capacity zero.

In view of (84) we have obtained the following result (Nakai [6]) anticipated at the end of 12C:

Theorem. *The set E of points of S for which* $\liminf B_k(a)$ *is infinite has vanishing capacity.*

16B. Again let f be an analytic mapping of R into S and set

$$E_n = \{a \mid a \in S, \nu(k, a) \leq n \text{ for all } k(\Omega)\}$$

with $n = 0, 1, \cdots$ and

$$E_\infty = \bigcup_{n=0}^{\infty} E_n.$$

In particular, E_0 is the set of all Picard points. The following statement is due to Nakai [6].

Theorem. *For arbitrary analytic mappings f with*

(85) $$\lim_{R_k \to R} \frac{C(k)}{k} = \infty$$

the set E_∞ has capacity zero. In other words, f covers S infinitely often except for a set of S of capacity zero.

Proof. We have only to show that E_n has capacity zero. Suppose this were not so. As in 16A there would then exist a nonzero measure μ such that $S_\mu \subset E_n$,

$$\tilde{s}^\mu \leq m$$

on S, and we would have

$$\int \tilde{B}_k(a)\, d\mu(a) = O(1) \tag{86}$$

for all R_k. We replace $B(k, a)$ by $\tilde{B}_k(a)$ in (60) and obtain

$$A(k, a) + \tilde{B}_k(a) = C(k) + O(k). \tag{87}$$

By (86) integration of both sides with respect to $d\mu(a)$ yields

$$\int A(k, a)\, d\mu(a) = C(k)\mu(S) + O(k). \tag{88}$$

By Fubini's theorem

$$\int A(k, a)\, d\mu(a) = 4\pi \int_0^k \left\{ \int_S \nu(h, a)\, d\mu(a) \right\} dh \leq 4\pi\mu(S)nk.$$

This together with (88) gives $C(k) = O(k)$, contradicting (85).

CHAPTER III

FUNCTIONS OF BOUNDED CHARACTERISTIC

A major part of the classical value distribution theory is based on the Poisson-Jensen formula. For the general theory of meromorphic functions f on Riemann surfaces R this is not a preferred method as even the first main theorem appears with a remainder that causes unnecessary complications. However, the approach has the advantage that $\log |f|$ on a regular subregion Ω can be decomposed. In the limit $\Omega \to R$ valuable information is thus obtained provided f has bounded characteristic. It is this class MB of meromorphic functions of bounded characteristic that will occupy us in the present chapter.

The chapter constitutes a natural complementation of Chapters I and II, where the results are meaningful only for mappings of unbounded characteristic.

In §1 we derive the decomposition theorem on Riemann surfaces; the proof is "direct" in that no use is made here of uniformization. In §2 we give a shorter proof employing uniformization and taking the corresponding classical theorem in the disk for granted. (For fundamentals on uniformization see, e.g., Ahlfors-Sario [1, p. 181].) We also relate the class O_{MB} of Riemann surfaces without nonconstant functions of bounded characteristic to other properties of function-theoretic significance.

It is possible to generalize Poisson's formula also to mappings between Riemann surfaces and to consider such mappings of bounded characteristic (Fuller [1]). Here we shall, however, always have the extended plane as the range surface.

The results in §1 are due to Sario [8], [11]; those in §2 to Heins [3], Parreau [2], and Rao [2].

§1. DECOMPOSITION

A meromorphic function of bounded characteristic in a disk is the quotient of two bounded analytic functions. This classical theorem can be extended to open Riemann surfaces R as follows. Consider the class MB of meromorphic functions f of bounded characteristic on R, defined in terms of capacity functions on subregions. Let L be the class of harmonic functions on R, regular except for logarithmic singularities with integral coefficients. Then $f \in MB$ if and only if $\log |f|$ is the difference of two positive functions in L. In this section we shall construct these functions on the surface R itself rather than on the disk of uniformization.

If $\log |f|$ is regular at the singularity of the capacity functions, then the classical reasoning (e.g., Nevanlinna [22]) can be generalized without difficulty. In the general case we introduce the extended class M_e of locally meromorphic functions e^{u+iu^*}, $u \in L$, with single-valued moduli. This class seems to be of interest in its own right.

1. Generalization of Jensen's formula

1A. Let $\bar{\Omega}$ be a compact bordered Riemann surface with border β_Ω, and let p denote the capacity function in $\bar{\Omega}$ with pole at a given point $z_0 \in \Omega$. By definition, $p(z) - \log |z - z_0| \to 0$ as $z \to z_0$, and $p(z) = k_\Omega =$ const. on β_Ω. (Here we use a slightly different normalization than in I.12A.)

Given a continuous real-valued function f on β_Ω, the solution v of the Dirichlet problem can be expressed in the form $v(z_0) = (2\pi)^{-1} \int_{\beta_\Omega} f\, dp^*$. To see this let α be a level line $p = c$ near z_0, oriented to leave z_0 to its left. Then by Green's formula $\int_{\beta_\Omega - \alpha} v\, dp^* - p\, dv^* = 0$, and the statement follows on letting $c \to -\infty$.

There is a simple relation between p with singularity at z_0 and the Green's function with singularity at $t \in \Omega$: $g(z_0, t) = k_\Omega - p(t)$. For the proof let δ be a level line $g(z, t) = c_1$ near t, encircling t counterclockwise. Then $\int_{\beta_\Omega - \alpha - \delta} g\, dp^* - p\, dg^* = 0$, and the statement is obtained in the limit as $c \to -\infty$, $c_1 \to \infty$. It is also a direct consequence of the well-known symmetry $g(z_0, t) = g(t, z_0)$.

1B. Let R be an arbitrary open Riemann surface. We consider the class $L = L(R)$ of functions u on R, harmonic except for logarithmic singularities $\lambda_i \log |z - z_i|$ at z_i, $i = 1, 2, \cdots$, with integral coefficients λ_i. By definition the class $M_e = M_e(R)$ consists of locally meromorphic functions

$$f = e^{u+iu^*} \quad \text{with } u \in L. \tag{1}$$

The conjugate function u^* has periods around z_i and along some cycles in R. Every branch of f is meromorphic, the branches differing by multiplicative constants d with $|d| = 1$. The modulus $|f|$ is single-valued throughout R. The class M_e contains the class M of (single-valued) meromorphic functions on R.

1C. Given $z_0 \in R$ let Ω be a regular subregion containing z_0 and with boundary β_Ω. Denote by a_μ, b_ν the zeros and poles of a given $f \in M_e$ on R.

We first assume that $f(z_0) \neq 0, \infty$, and that no a_μ, b_ν is on β_Ω. Consider on $\bar{\Omega}$ the function

$$v(z) = \log |f(z)| + \sum_{a_\mu \in \Omega} g(z, a_\mu) - \sum_{b_\nu \in \Omega} g(z, b_\nu),$$

where each a_μ, b_ν is taken as many times as indicated by its multiplicity. Clearly v is harmonic on $\bar{\Omega}$, and

$$(2) \quad \log |f(z_0)| = \frac{1}{2\pi} \int_{\beta_\Omega} \log |f(z)| \, dp^* - \sum_{a_\mu \in \Omega} (k_\Omega - p(a_\mu)) + \sum_{b_\nu \in \Omega} (k_\Omega - p(b_\nu)).$$

If an a_μ or b_ν is on β_Ω, we first apply (2) to a slightly smaller region $\Omega_{k_\Omega - \varepsilon} \subset \Omega$ bounded by the level line $p = k_\Omega - \varepsilon$, and then let $\varepsilon \to 0$. Since all terms in the equation are continuous in ε, (2) remains valid for Ω.

1D. Suppose now $f(z_0) = 0$ or ∞. A branch of f near z_0 then has the Laurent expansion

$$f(z) = c_\lambda (z - z_0)^\lambda + c_{\lambda+1}(z - z_0)^{\lambda+1} + \cdots,$$

and the other branches are obtained on multiplying by constants e^{ir}, with r real. The same is true of the branches of the function

$$\varphi(z) = e^{p(z) + ip^*(z)} \in M_e$$

in $\bar{\Omega}$ and of

$$\psi(z) = f(z) \cdot \varphi(z)^{-\lambda} = c_\lambda + \varepsilon(z - z_0) \in M_e,$$

where $\varepsilon(z - z_0) \to 0$ as $z \to z_0$. On applying (2) to ψ we obtain

$$\log |c_\lambda| = \frac{1}{2\pi} \int_{\beta_\Omega} \log |f| \, dp^* - \lambda k_\Omega - \sum{}' (k_\Omega - p(a_\mu)) + \sum{}' (k_\Omega - p(b_\nu)),$$

the sums $\sum'$ being extended over points in $\bar{\Omega} - z_0$.

1E. For $-\infty < h \leq k_\Omega$ consider the region $\Omega_h \subset \Omega$ bounded by the level line $p = h$. Let $n(h, a)$ be the number of a-points, $a = 0$ or ∞, of f in $\bar{\Omega}_h$, counted with their multiplicities. It is understood that $n(-\infty, a)$ is the multiplicity (≥ 0) of the a-point at z_0. Then

$$\sum{}' (k_\Omega - p(f^{-1}(a))) = \int_{-\infty}^{k_\Omega} (k_\Omega - h) \, d(n(h, a) - n(-\infty, a))$$

$$= \int_{-\infty}^{k_\Omega} (n(h, a) - n(-\infty, a)) \, dh,$$

and $\lambda = n(-\infty, 0) - n(-\infty, \infty)$. We set

$$(3) \qquad N(h, a) = \int_{-\infty}^{h} (n(h, a) - n(-\infty, a)) \, dh + n(-\infty, a) h$$

and also use the notations $N(\Omega, a) = N(k_\Omega, a)$, $N(\Omega, f) = N(\Omega, \infty)$. We have obtained the following auxiliary result (Sario [11]):

Jensen's formula on Riemann surfaces. *For a locally meromorphic function $f \in M_e$ with single-valued modulus on an arbitrary open Riemann surface R,*

$$\log |c_\lambda| = \frac{1}{2\pi} \int_{\beta_\Omega} \log |f| \, dp^* + N(\Omega, f) - N\left(\Omega, \frac{1}{f}\right). \tag{4}$$

1F. Using $\overset{+}{\log} |f| = \max (0, \log |f|)$ we set

$$m(\Omega, f) = \frac{1}{2\pi} \int_{\beta_\Omega} \overset{+}{\log} |f| \, dp^* \tag{5}$$

and see that

$$\frac{1}{2\pi} \int_{\beta_\Omega} \log |f| \, dp^* = m(\Omega, f) - m\left(\Omega, \frac{1}{f}\right).$$

The counterpart of Nevanlinna's *characteristic function* is

$$T(\Omega) = T(\Omega, f) = m(\Omega, f) + N(\Omega, f), \tag{6}$$

and Jensen's formula (4) takes the form

$$\log |c_\lambda| = T(\Omega, f) - T\left(\Omega, \frac{1}{f}\right) \tag{7}$$

for all $f \in M_e$.

1G. We shall now consider differences $f-a$, and we therefore restrict our attention to the class $M \subset M_e$ of single-valued meromorphic functions f on R.

For $a \neq \infty$ we define the counterpart of Nevanlinna's *proximity function* as

$$m(\Omega, a) = m\left(\Omega, \frac{1}{f-a}\right) = \frac{1}{2\pi} \int_{\beta_\Omega} \overset{+}{\log} \frac{1}{|f-a|} \, dp^*, \tag{8}$$

and the counterpart of the *counting function* as

$$N(\Omega, a) = N\left(\Omega, \frac{1}{f-a}\right) = \int_{-\infty}^{k_\Omega} (n(h, a) - n(-\infty, a)) \, dh + n(-\infty, a) k_\Omega. \tag{9}$$

For $a = \infty$ the definitions were given by (5) and (3).

We apply Jensen's formula (7) to $f-a$. Clearly $N(\Omega, f-a) = N(\Omega, f)$, while the inequality $\log^+ (p_1 + p_2) \leq \log^+ p_1 + \log^+ p_2 + \log 2$ for arbitrary numbers $p_1, p_2 > 0$ yields

$$m(\Omega, f-a) \leq m(\Omega, f) + \overset{+}{\log} |a| + \log 2.$$

We can now state the following extension to Riemann surfaces of Nevanlinna's form of the first main theorem:

Theorem. *For a meromorphic function on an arbitrary Riemann surface R*

$$m(\Omega, a) + N(\Omega, a) = T(\Omega) + \varphi(a), \tag{10}$$

where $|\varphi(a)| \leq \overset{+}{\log} |a| + \log 2 + |\log |c||$, *c being the first Laurent coefficient of* $f-a$ *at* z_0.

The theorem simply states that in the Poisson-Jensen decomposition of $\log |f(z)-a|$ into positive and negative harmonic functions on $\bar{\Omega}$ their values at z_0 must add up to $\log |f(z_0)-a|$. The value of the positive component of $\log |f(z)|$ at z_0 is the characteristic $T(\Omega)$.

2. Decomposition theorem

2A. We introduce the *class MB (or* M_eB*) of functions f in M (or* M_e*) with bounded characteristics*

$$T(\Omega) = O(1). \tag{11}$$

Explicitly, one requires the existence of a bound $M < \infty$ independent of Ω such that $T(\Omega) < M$ for all $\Omega \subset R$. That (11) is independent of z_0 will be a consequence of a decomposition theorem which we proceed to establish.

We continue considering arbitrary open Riemann surfaces R. Let LP be the subclass consisting of all positive functions in the class L defined in 1B and the constant function 0. We are ready to state (Sario [8]):

Theorem. *A necessary and sufficient condition for* $f \in M_eB$ *on R is that*

$$\log |f| = u - v, \tag{12}$$

where $u, v \in LP$.

The proof will be given in 2B to 2L. As a corollary we observe that a function $f \in M$ is in MB on R if and only if (12) holds.

2B. First we shall discuss in 2B to 2E the case $f(z_0) = 0$ or ∞.

Suppose $f \in M_eB$. We begin by showing that $R \notin O_G$. If $f(z_0) = \infty$, then

$$T(\Omega) \geq N(\Omega, f) \geq n(-\infty, \infty)k_\Omega \geq k_\Omega.$$

If $R \in O_G$, then $k_\Omega \to \infty$ as $\Omega \to R$ (App. I.1) and consequently $T(\Omega) \to \infty$, a contradiction. If $f(z_0) = 0$, then from Jensen's formula

$$T(\Omega, f) = T\left(\Omega, \frac{1}{f}\right) + O(1)$$

we have

$$T\left(\Omega, \frac{1}{f}\right) \geq N\left(\Omega, \frac{1}{f}\right) \geq n(-\infty, 0)k_\Omega \geq k_\Omega$$

and arrive at the same conclusion $R \notin O_G$.

On the other hand, if equality (12) is true, the existence of nonnegative superharmonic functions u, v, at least one of which is nonconstant, implies $R \notin O_G$ (App. I.4). Thus either condition of the theorem gives the hyperbolicity of R, and we may henceforth assume the existence of the Green's function $g(z, z_0)$ on R if $f(z_0)=0$ or ∞.

2C. Let n_0, n_∞ be the multiplicities of the zero or pole of f at z_0; then $\lambda = n_0 - n_\infty$. The functions

$$\varphi(z) = e^{\lambda(g(z,z_0)+ig^*(z,z_0))}, \qquad f_1(z) = f(z)\varphi(z)$$

belong to M_e, and we can reword the problem of characterizing M_eB:

Lemma. *A necessary and sufficient condition for $f \in M_eB$ is that $f_1 \in M_eB$.*

Proof. By definition,

$$T(\Omega, \varphi) = N(\Omega, \varphi) + m(\Omega, \varphi).$$

For $\lambda > 0$ we have trivially $N(\Omega, \varphi^{-1}) \equiv 0$, $m(\Omega, \varphi^{-1}) \equiv 0$, whence $T(\Omega, \varphi^{-1}) \equiv 0$, and it follows from Jensen's formula that $T(\Omega, \varphi) = O(1)$. If $\lambda < 0$, then $N(\Omega, \varphi) \equiv m(\Omega, \varphi) \equiv 0$, $T(\Omega, \varphi) \equiv 0$, and consequently $T(\Omega, \varphi^{-1}) = O(1)$. In both cases

$$T(\Omega, \varphi) = O(1), \qquad T(\Omega, \varphi^{-1}) = O(1).$$

The inequalities

$$T(\Omega, f) \leq T(\Omega, f_1) + T(\Omega, \varphi^{-1}) = T(\Omega, f_1) + O(1),$$
$$T(\Omega, f_1) \leq T(\Omega, f) + T(\Omega, \varphi) = T(\Omega, f) + O(1)$$

yield

$$T(\Omega, f) = T(\Omega, f_1) + O(1), \tag{13}$$

and the lemma follows.

2D. Reformulation of condition (12) is equally elementary:

Lemma. *A necessary and sufficient condition for*

$$\log|f| = u - v \tag{14}$$

with u, $v \in LP$ is that

$$\log|f_1| = u_1 - v_1 \tag{15}$$

with u_1, $v_1 \in LP$.

Proof. We know that

$$\log|f_1| = \log|f| + \lambda g = \log|f| + (n_0 - n_\infty)g.$$

If (14) is true, then

$$\log |f_1| = (u + n_0 g) - (v + n_\infty g)$$

and (15) follows. The converse is seen similarly.

2E. We conclude that Theorem 2A will be proved for f with $f(z_0)=0$ or ∞ if we establish it for f_1. Explicitly, we are to show that $f_1 \in M_e B$ if and only if $\log |f_1| = u_1 - v_1$, $u_1, v_1 \in LP$.

2F. Let $p_{\Omega z}$ be the capacity function in $\bar{\Omega}$ with pole at z. For a harmonic function h on $\bar{\Omega}$, $h(z) = (2\pi)^{-1} \int_{\beta_\Omega} h \, dp^*_{\Omega z}$.

Denote by a_μ, b_ν the zeros and poles of f in R. Those in $R - z_0$ are the zeros and poles of f_1 in R. Suppose first there is no a_μ, b_ν on β_Ω. Then the function

$$h(z) = \log |f_1(z)| + \sum_{a_\mu \in \Omega - z_0} g_\Omega(z, a_\mu) - \sum_{b_\nu \in \Omega - z_0} g_\Omega(z, b_\nu)$$

is harmonic on $\bar{\Omega}$. Throughout this section the zeros and poles are counted with their multiplicities. We set

(16) $$x_\Omega(z, f_1) = \frac{1}{2\pi} \int_{\beta_\Omega} \overset{+}{\log} |f_1| dp^*_{\Omega z},$$

(17) $$y_\Omega(z, f_1) = \sum_{b_\nu \in \Omega - z_0} g_\Omega(z, b_\nu),$$

and

(18) $$u_\Omega(z, f_1) = x_\Omega(z, f_1) + y_\Omega(z, f_1).$$

Then

(19) $$\log |f_1(z)| = u_\Omega(z, f_1) - u_\Omega(z, f_1^{-1}).$$

Since all terms are continuous in Ω, the equation remains valid if there are zeros or poles of f on β_Ω.

We observe that $x_\Omega(z_0, f_1) = m(\Omega, f_1)$ and $y_\Omega(z_0, f_1) = N(\Omega, f_1)$. Here we shall only make use of the consequence

(20) $$u_\Omega(z_0, f_1) = T(\Omega, f_1).$$

2G. We shall show next:

Lemma. *For* $\Omega_0 \subset \Omega$

(21) $$u_{\Omega_0}(z, f_1) \leq u_\Omega(z, f_1),$$

(21)′ $$u_{\Omega_0}(z, f_1^{-1}) \leq u_\Omega(z, f_1^{-1}).$$

Proof. By (19), $\overset{+}{\log}\,|f_1(z)| \le u_\Omega(z, f_1)$ for every Ω. It follows that

$$x_{\Omega_0}(z, f_1) \le \frac{1}{2\pi}\int_{\beta_{\Omega_0}} u_\Omega(t, f_1)\, dp^*_{\Omega_0 z}$$

$$= \frac{1}{2\pi}\int_{\beta_{\Omega_0}} (u_\Omega(t, f_1) - y_{\Omega_0}(t, f_1))\, dp^*_{\Omega_0 z}$$

$$= u_\Omega(z, f_1) - y_{\Omega_0}(z, f_1),$$

because this difference is regular harmonic in Ω_0. We have reached statement (21), and inequality (21)′ follows in the same fashion.

2H. From (20) and (21) we infer that $T(\Omega, f_1)$ increases with Ω. We can set

$$T(R, f_1) = \lim_{\Omega \to R} T(\Omega, f_1) \tag{22}$$

and use alternatively the notations $T(\Omega) = O(1)$ and $T(R) < \infty$.

2I. The convergence of u_Ω can now be established:

Lemma. *If $T(R, f_1) < \infty$, then the functions*

$$u(z, f_1) = \lim_{\Omega \to R} u_\Omega(z, f_1) \quad \text{and} \quad u(z, f_1^{-1}) = \lim_{\Omega \to R} u_\Omega(z, f_1^{-1}) \tag{23}$$

are positive harmonic on R except for logarithmic poles of $u(z, f_1)$ at the $b_\nu \in R - z_0$ and those of $u(z, f_1^{-1})$ at the $a_\mu \in R - z_0$.

Proof. By Harnack's principle the first limit in (23) is either identically infinite or else harmonic on $R - \{b_\nu\}$. That the latter alternative occurs is a consequence of $\lim_{\Omega \to R} u_\Omega(z_0, f_1) = T(R, f_1)$. The statement for $u_\Omega(z, f_1^{-1})$ follows similarly from $u_\Omega(z_0, f_1^{-1}) = T(\Omega, f_1^{-1}) = T(\Omega, f_1) + O(1)$.

2J. On combining the lemma with (19) we see that $f_1 \in M_e B$ has the asserted representation

$$\log|f_1(z)| = u(z, f_1) - u(z, f_1^{-1}) \tag{24}$$

with the u-functions in LP. It remains to establish the converse.

2K. Suppose

$$\log|f_1(z)| = u_1(z) - v_1(z), \tag{25}$$

where u_1, $v_1 \in LP$. The positive logarithmic poles of $u_\Omega(z, f_1)$ are those of $\log|f_1(z)|$ in Ω, hence among those of $u_1(z)$. Consequently $u_1(z) - u_\Omega(z, f_1)$ is superharmonic in Ω and its minimum on $\bar{\Omega}$ is reached on β_Ω, where $u_1(z) - u_\Omega(z, f_1) = u_1(z) - \log^+|f_1(z)| \ge 0$. One infers that $u_1(z) \ge u_\Omega(z, f_1)$ in $\bar{\Omega}$. At z_0 this means

$$T(\Omega, f_1) = u_\Omega(z_0, f_1) \le u_1(z_0).$$

For the case $u_1(z_0) < \infty$ the proof is complete.

2L. If $u_1(z_0)=\infty$, then $u_1(z)+\lambda_1 \log|z-z_0|$ is harmonic at z_0 for some positive integer λ_1. We set

$$f_2 = f_1 \cdot e^{-\lambda_1(g+ig^*)} \in M_e, \tag{26}$$

where $g=g(z, z_0)$, and obtain

$$\log|f_2| = \log|f_1|-\lambda_1 g = (u_1 - \lambda_1 g)-v_1.$$

The function $u_1-\lambda_1 g_\Omega$ with $g_\Omega=g_\Omega(z, z_0)$ is superharmonic on Ω, hence its minimum on $\bar{\Omega}$ is taken on β_Ω, where $u_1-\lambda_1 g_\Omega=u_1\geq 0$. From $u_1\geq\lambda_1 g_\Omega$ on Ω it follows that

$$u_1-\lambda_1 g = \lim_{\Omega\to R}(u_1-\lambda_1 g_\Omega) \geq 0$$

on R. On setting $u_2=u_1-\lambda_1 g$ and $v_2=v_1$ we obtain

$$\log|f_2| = u_2-v_2 \tag{27}$$

with $u_2, v_2 \in LP$.

The positive logarithmic poles of $u_\Omega(z, f_2)$ are those of $\log|f_2|$ on Ω, hence among those of u_2. The minimum of the superharmonic function $u_2(z)-u_\Omega(z, f_2)$ on $\bar{\Omega}$ is taken on β_Ω, where it is $\min_{\beta_\Omega}(u_2-\log^+|f_2|)\geq 0$. We infer that

$$T(\Omega, f_2) = u_\Omega(z_0, f_2) \leq u_2(z_0) < \infty,$$

i.e., $T(\Omega, f_2)=O(1)$. The reasoning leading to (13) yields

$$T(\Omega, f_1) = T(\Omega, f_2)+O(1),$$

and consequently $T(\Omega, f_1)=O(1)$.

We have shown that (25) implies $T(R, f_1)<\infty$, and the proof of Theorem 2A is complete.

2M. As an immediate consequence we see that the property $T(\Omega, f)=O(1)$ and thus the class M_eB is independent of z_0.

3. Extremal decompositions

3A. Consider an arbitrary $f \in M_e$. We make no restrictive assumptions on $f(z_0)$ and form

$$x_\Omega(z, f) = \frac{1}{2\pi}\int_{\beta_\Omega} \overset{+}{\log}|f|\, dp^*_{\Omega z},$$

$$y_\Omega(z, f) = \sum_{b_\nu\in\Omega} g_\Omega(z, b_\nu),$$

$$u_\Omega(z,f) = x_\Omega(z, f)+y_\Omega(z, f).$$

It is seen as in 2G that u_Ω increases with Ω and that

$$u(z,f) = \lim_{\Omega\to R} u_\Omega(z,f) \tag{28}$$

is either identically infinite or else positive harmonic on R except for logarithmic poles b_ν. The same is true of

$$u(z,f^{-1}) = \lim_{\Omega\to R} u_\Omega(z,f^{-1}) \tag{29}$$

with singularities a_μ.

The functions (28) and (29) will now be shown to be extremal in all decompositions (12):

Theorem. *If there is a decomposition*

$$\log|f(z)| = u_1(z) - u_2(z) \tag{30}$$

with u_1, $u_2 \in LP$, then also

$$\log|f(z)| = u(z,f) - u(z,f^{-1}). \tag{31}$$

Moreover,

$$u(z,f) \le u_1(z) \quad \textit{and} \quad u(z,f^{-1}) \le u_2(z). \tag{32}$$

Proof. One observes that the positive logarithmic poles of $u_\Omega(z,f)$ are those of $\log|f(z)|$ in Ω, hence among those of $u_1(z)$ in Ω. The superharmonic function $u_1(z) - u_\Omega(z,f)$ on Ω dominates $\min_{\beta_\Omega}(u_1(z) - \log^+|f(z)|) \ge 0$ and we find that $u_1(z) - u(z,f) = \lim_{\Omega\to R}(u_1(z) - u_\Omega(z,f)) \ge 0$ on R. Similarly the superharmonic function $u_2(z) - u_\Omega(z,f^{-1}) \ge 0$ on Ω, and $u_2(z) \ge u(z,f^{-1})$ on R. By virtue of Harnack's principle equality (31) then follows on letting $\Omega \to R$ in

$$\log|f(z)| = u_\Omega(z,f) - u_\Omega(z,f^{-1}).$$

3B. The extremal functions $u(z,f)$, $u(z,f^{-1})$ can in turn be decomposed (Sario [8]):

Theorem. *A function f on R belongs to M_eB if and only if*

$$\log|f| = (x(z,f) + y(z,f)) - (x(z,f^{-1}) + y(z,f^{-1})), \tag{33}$$

where the functions $x \ge 0$ are regular harmonic and the functions $y \ge 0$ have the representations

$$y(z,f) = \sum g(z,b_\nu), \qquad y(z,f^{-1}) = \sum g(z,a_\mu). \tag{34}$$

Here the sums are extended over all poles b_ν and all zeros a_μ of f in R, respectively, each counted with its multiplicity.

3C. Suppose indeed that $f \in M_eB$. It is evident from the maximum principle that $y_{\Omega_0}(z,f) \le y_\Omega(z,f)$ for $\Omega_0 \subset \Omega$. We know that $\log|f| =$

u_1-u_2, $u_1, u_2 \in LP$, and the superharmonic function $u_1(z)-y_\Omega(z,f)$ on Ω majorizes $\min_{\beta_\Omega} u_1 \geq 0$. Hence $y_\Omega(z,f) \leq u_1(z)$ on Ω and, by Harnack's principle,

(35) $$y(z,f) = \lim_{\Omega \to R} y_\Omega(z,f)$$

is positive harmonic on R except for logarithmic poles b_ν. An analogous reasoning shows that

(36) $$y(z,f^{-1}) = \lim_{\Omega \to R} y_\Omega(z,f^{-1})$$

is positive harmonic on $R-\{a_\mu\}$.

3D. To prove (34) we must show that

(37) $$\lim_{\Omega \to R} \sum_{b_\nu \in \Omega} g_\Omega(z,b_\nu) = \sum_{b_\nu \in R} g(z,b_\nu)$$

and similarly for $\sum g(z,a_\mu)$. First,

$$\sum_{b_\nu \in \Omega} g_\Omega(z,b_\nu) \leq \sum_{b_\nu \in \Omega} g(z,b_\nu) \leq \sum_{b_\nu \in R} g(z,b_\nu),$$

and we have

$$\limsup_{\Omega \to R} \sum_{b_\nu \in \Omega} g_\Omega(z,b_\nu) \leq \sum_{b_\nu \in R} g(z,b_\nu).$$

Second, for $\Omega_0 \subset \Omega$

$$\sum_{b_\nu \in \Omega_0} g(z,b_\nu) = \lim_{\Omega \to R} \sum_{b_\nu \in \Omega_0} g_\Omega(z,b_\nu) \leq \liminf_{\Omega \to R} \sum_{b_\nu \in \Omega} g_\Omega(z,b_\nu)$$

and a fortiori

$$\sum_{b_\nu \in R} g(z,b_\nu) = \lim_{\Omega_0 \to R} \sum_{b_\nu \in \Omega_0} g(z,b_\nu) \leq \liminf_{\Omega \to R} \sum_{b_\nu \in \Omega} g_\Omega(z,b_\nu).$$

Statement (37) follows.

3E. The convergence of $x_\Omega(z,f)$ is obtained at once from $x_\Omega(z,f) = u_\Omega(z,f)-y_\Omega(z,f)$, and the limiting function is

(38) $$x(z,f) = u(z,f)-y(z,f).$$

The limit $x(z,f^{-1})$ of $x_\Omega(z,f^{-1})$ is obtained in the same way. Both limits are obviously positive and regular harmonic on R.

The necessity of (33) for $f \in M_e B$ has thus been established. The sufficiency is a corollary of the main Theorem 2A.

4. Consequences

4A. We close this section with a number of immediate consequences. The proofs are left to the reader (cf. Sario [8]).

If only the x-terms in (33) are considered, the following simpler but less accurate result is obtained from Theorem 3B:

Corollary 1. *If $f \in M_eB$ on R, then*

$$\lim_{\Omega \to R} \int_{\beta_\Omega} |\log |f|| \, dp_\Omega^* < \infty \tag{39}$$

for any z_0.

Here p_Ω signifies, as before, the capacity function on Ω with its pole at z_0.

A consideration of the y-terms in (33) gives:

Corollary 2. *Suppose $f \in M_eB$. Then the sum $\sum g(z, z_i)$, with z_i ranging over all poles and zeros of f, is harmonic on $R - \{a_\mu\} - \{b_\nu\}$.*

For a sufficient condition the first terms of both x- and y-parts in (33) must be taken into account:

Corollary 3. *If for some $z_0 \in R$*

$$\int_{\beta_\Omega} \overset{+}{\log} |f| \, dp_\Omega^* = O(1) \tag{40}$$

and

$$\sum_{b_\nu \in R} g(z, b_\nu) < \infty \quad \text{on} \quad R - \{b_\nu\}, \tag{41}$$

then $f \in M_eB$.

A fortiori

$$\lim_{\Omega \to R} \int_{\beta_\Omega} |\log |f|| \, dp_\Omega^* < \infty \tag{42}$$

and

$$\sum_{a_\mu \in R} g(z, a_\mu) < \infty \quad \text{on} \quad R - \{a_\mu\} \tag{43}$$

as well.

Another sufficient condition for $f \in M_eB$ is, of course, that

$$\int_{\beta_\Omega} \overset{+}{\log} |f^{-1}| \, dp_\Omega^*$$

is bounded and $\sum g(z, a_\mu) < \infty$ in $R - \{a_\mu\}$.

4B. For "entire" functions in M_eB the conditions simplify. Let E_eB be the class of such functions, characterized by $f(z) \neq \infty$ on R.

Corollary 4. *A necessary and sufficient condition for $f \in E_e B$ on R is that*

$$\int_{\beta_\Omega} \overset{+}{\log} |f| \, dp_\Omega^* = O(1). \tag{44}$$

Consider the class H of regular harmonic functions h on R and let HP be the subclass of nonnegative functions. Set $h^+ = \max(0, h)$.

Corollary 5. *A harmonic function h on R has a decomposition*

$$h = u_1 - u_2, \qquad u_1, u_2 \in HP \tag{45}$$

if and only if for some z_0

$$\int_{\beta_\Omega} h^+ \, dp_\Omega^* = O(1), \tag{46}$$

or, equivalently,

$$\lim_{\Omega \to R} \int_{\beta_\Omega} |h| \, dp_\Omega^* < \infty. \tag{47}$$

It is known that functions harmonic in the interior R of a compact bordered Riemann surface and with property (47) have a Poisson-Stieltjes representation (e.g., Rodin [1]). For further interesting results in this direction, see Rao [1].

4C. It is clear that theorems on $\log |f|$ can also be expressed directly in terms of $|f|$. Theorem 2A, e.g., takes the following form:

Corollary 6. *$f \in M_e B$ if and only if*

$$|f| = \left| \frac{\eta(z, f)}{\eta(z, f^{-1})} \right|, \tag{48}$$

where $\eta \in M_e B$ and $|\eta| < 1$ on R.

The counterpart of Theorem 3B is the following:

Corollary 7. *$f \in M_e B$ if and only if*

$$|f| = \left| \frac{\varphi(z, f)\psi(z, f)}{\varphi(z, f^{-1})\psi(z, f^{-1})} \right|, \tag{49}$$

where $\varphi, \psi \in M_e B$ and $\varphi \neq 0$ on R, $|\varphi| < 1$, $|\psi| < 1$.

§2. THE CLASS O_{MB}

In the preceding section we carried out the decompositions directly on the Riemann surface R. In the special case of the disk our results give the classical decomposition theorems. We now reverse the viewpoint by taking these classical theorems for granted and, using uniformization, derive the decompositions on Riemann surfaces.

Our reasoning will be incorporated into a discussion of the classes O_{MB} and O_{M_eB} of Riemann surfaces on which there are no nonconstant functions in MB and M_eB, respectively. In passing we also consider the classes O_{EB} and O_{E_eB} determined by the entire functions in MB and M_eB. The problem here is to arrange these four classes in the general classification scheme of Riemann surfaces and to consider their other function-theoretic properties.

5. Preliminaries

5A. We shall again denote by f a meromorphic function on a Riemann surface R, by ζ the variable in the range sphere, and by z_0 the singular point of the capacity function.

Unless otherwise stated we shall consider only nonconstant meromorphic functions f. We fix the point z_0 and denote by $T(\Omega)=T(\Omega, f)$ the characteristic of f with respect to z_0 and a regular region Ω which contains z_0.

The inclusion relations

$$O_{M_eB} \subset O_{MB} \subset O_{EB},$$
$$O_{M_eB} \subset O_{E_eB} \subset O_{EB} \tag{50}$$

are trivially true, and the characterization of the smallest class is immediate:

Theorem. *All functions in M_eB on R reduce to constants if and only if R is parabolic*, i.e.,

$$O_G = O_{M_eB}. \tag{51}$$

Proof. If $R \notin O_G$, there is a Green's function $g(z, z_0)$, and, by Theorem 2A,

$$f = e^{-g-ig^*} \in M_eB. \tag{52}$$

Conversely, if there is a nonconstant $f \in M_eB$ on R, then $\log |f| = u_1 - u_2$, where $u_i \in LP$ is superharmonic and at least one of u_1 and u_2 is nonconstant. This means that $R \notin O_G$ (App. I.4.). The same proof gives $O_G = O_{E_eB}$.

5B. By the preceding theorem every M_e-function on a parabolic R has unbounded characteristic. Even more can be said of M-functions by comparing $T(\Omega)$ with k_Ω:

Theorem. *On $R \in O_{MB}$ the characteristic $T(\Omega)$ of any nonconstant $f \in M$ tends so rapidly to infinity that*

$$\liminf_{\Omega \to R} \frac{T(\Omega)}{k_\Omega} \geq 1. \tag{53}$$

Proof. Let $f(z_0) = a$. The counting function of f for a is, by definition,

$$N(\Omega, a) = \int_{-\infty}^{k_\Omega} (n(h, a) - n(-\infty, a))\, dh + n(-\infty, a) k_\Omega,$$

where $n(h, a)$ is the number of a-points of f in the set $\bar{\Omega}_h = \{z \mid p_\Omega(z) \leq h \leq k_\Omega\}$. We infer from the first main theorem (10) that

$$T(\Omega) + O(1) > N(\Omega, a) \geq n(-\infty, a) k_\Omega,$$

and (53) follows.

5C. For later reference we insert here a proof of Frostman's [3] formula and as a corollary the formula of Cartan [4], both extended to Riemann surfaces. Suppose that E is a compact subset of the complex plane which does not contain the point $f(z_0) \neq \infty$. Applying Jensen's formula (4) to the function $f - \zeta$ with $\zeta \in E$ we obtain

$$\log |f(z_0) - \zeta| = \frac{1}{2\pi} \int_{\beta_\Omega} \log |f(z) - \zeta|\, dp^* + N(\Omega, \infty) - N(\Omega, \zeta).$$

If μ is a positive measure on E with total measure unity, then we integrate the above relation with respect to $d\mu(\zeta)$ over E and set

$$u(\xi) = \int_E \log \frac{1}{|\xi - \zeta|}\, d\mu(\zeta).$$

The result is the *extension to Riemann surfaces of the formula of Frostman*:

$$-u(f(z_0)) = -\frac{1}{2\pi} \int_{\beta_\Omega} u(f(z))\, dp^* + N(\Omega, \infty) - \int_E N(\Omega, \zeta)\, d\mu(\zeta). \tag{54}$$

5D. We can specialize this equation by taking E to be the unit circle $|\zeta| = 1$, $\zeta = e^{i\vartheta}$ and setting $d\mu(\zeta) = (2\pi)^{-1} d\vartheta$. This yields

$$-u(\xi) = \frac{1}{2\pi} \int_0^{2\pi} \log |e^{i\vartheta} - \xi|\, d\vartheta.$$

If $|\xi| > 1$, then $\log |\zeta - \xi|$ is harmonic for $|\zeta| \leq 1$, and we infer from the Gauss mean value theorem that $-u(\xi) = \log |0 - \xi| = \log |\xi|$, or equivalently

$$-u(\xi) = \overset{+}{\log} |\xi|. \tag{55}$$

On the other hand, if $|\xi| < 1$ we observe that $\log |e^{i\vartheta} - \xi| = \log |1 - \xi e^{-i\vartheta}|$. Since $\log |1 - \bar{\zeta}\xi|$ is harmonic for $|\zeta| \leq 1$, we find as before that

$$-u(\xi) = \log |1 - 0 \cdot \xi| = \overset{+}{\log} |\xi|.$$

Since (55) is thus valid for $|\xi| \lessgtr 1$ it holds by continuity for $|\xi| = 1$.

On substituting in (54) we obtain the *generalization to Riemann surfaces of Cartan's formula*:

$$T(\Omega, f) = \overset{+}{\log} |f(z_0)| + \frac{1}{2\pi} \int_0^{2\pi} N(\Omega, e^{i\vartheta})\, d\vartheta. \tag{56}$$

This again shows that $T(\Omega, f)$ increases with Ω.

In terms of this evaluation one can also estimate

$$T(\Omega, f_1 f_2) \leq T(\Omega, f_1) + T(\Omega, f_2), \tag{57}$$

and

$$T(\Omega, f_1 + \cdots + f_q) \leq \sum_1^q T(\Omega, f_i) + \log q. \tag{58}$$

5E. We note that *the functions in MB constitute a field*. In fact, if $f_1, f_2 \in MB$, then by (57), (58) we have $f_1 + f_2, f_1 f_2 \in MB$, and Jensen's formula (7) shows that $f \in MB$ implies $f^{-1} \in MB$. For later reference we also observe that the first main theorem (10) gives $(f - \zeta)^{-1} \in MB$ if $f \in MB$, for any complex number ζ.

The problem of characterizing O_{MB} is: When does the field MB reduce to the field of constants?

6. Characterization of O_{MB}

6A. If the surface R is not a member of the class O_G of parabolic Riemann surfaces, then for a point a of R we consider the class of all positive superharmonic functions on R, with at least the singularity $-\log |z - a|$ at a. It is known that this class is not empty and that it has a smallest member $g(z, a) = g_R(z, a)$, the Green's function on R (App. I.5).

6B. We shall assume that $f \in MB$ on R, i.e., $T(\Omega, f) = O(1)$. We infer from the definition of the characteristic that

$$N(\Omega, f) = \sum_{\substack{f(a) = \infty \\ a \in \Omega - z_0}} \nu(f, a) g_\Omega(z_0, a) + n(-\infty, \infty) k_\Omega \leq M < \infty, \tag{59}$$

where $\nu(f, a)$ indicates the multiplicity of the pole of f at a. On $\bar{\Omega}$ we consider the function

$$y_\Omega(z, f) = \sum \nu(f, a) g_\Omega(z, a), \tag{60}$$

where the summation is over all poles $a \neq z_0$ of f in Ω.

As Ω increases more terms are added to the sum in (60). Since by the minimum principle the Green's functions increase with Ω we have $y_\Omega(z, f) \leq y_{\Omega'}(z, f)$ for $\Omega \subset \Omega'$ and $z \in \Omega$. By Harnack's principle and relations (59), (60), y_Ω tends to a positive harmonic function

$$y(z, f) = \lim_{\Omega \to R} y_\Omega(z, f)$$

in $R - f^{-1}(\infty)$. It is clear that $y(z, f)$ has positive logarithmic singularities with coefficients $\nu(f, a)$ at the poles $a \neq z_0$ of f. For some ζ the function $(f-\zeta)^{-1}$ has poles at points other than z_0 on R. Since this function is in MB if and only if f is (cf. 5E), we deduce that for some ζ, $y(z, (f-\zeta)^{-1})$ is a nonconstant positive superharmonic function. Consequently we find again the inclusion relation

$$O_G \subset O_{MB}. \tag{61}$$

6C. For any ζ of the extended plane we set

$$G(z, \zeta, f) = \sum_{f(a)=\zeta} \nu(f, a) g(z, a), \tag{62}$$

where $\nu(f, a)$ is the multiplicity of the ζ-point at a, and $g(z, a)$ is a Green's function on R. If $f(z_0) \neq \zeta$, then we shall show that

$$y(z, (f-\zeta)^{-1}) = G(z, \zeta, f) \tag{63}$$

and if $f(z_0) = \zeta$, then

$$\nu(f, z_0) g(z, z_0) + y(z, (f-\zeta)^{-1}) = G(z, \zeta, f). \tag{64}$$

In fact,

$$\begin{aligned} y_\Omega(z, (f-\zeta)^{-1}) &= \sum_{\substack{f(a)=\zeta \\ a \in \Omega - z_0}} \nu(f, a) g_\Omega(a, z) \\ &\leq \sum_{\substack{f(a)=\zeta \\ a \in \Omega - z_0}} \nu(f, a) g(a, z) \leq G'(z, \zeta, f), \end{aligned}$$

where the prime in G' indicates that the sum in (62) is extended over all $a \neq z_0$. It follows that

$$y(z, (f-\zeta)^{-1}) \leq G'(z, \zeta, f). \tag{65}$$

On the other hand, for a fixed Ω_0

$$\begin{aligned} \sum_{\substack{f(a)=\zeta \\ a \in \Omega_0 - z_0}} \nu(f, a) g(z, a) &= \lim_{\Omega \to R} \sum_{\substack{f(a)=\zeta \\ a \in \Omega_0 - z_0}} \nu(f, a) g_\Omega(z, a) \\ &\leq \lim_{\Omega \to R} y_\Omega(z, (f-\zeta)^{-1}) = y(z, (f-\zeta)^{-1}). \end{aligned}$$

Since this holds for every Ω_0, we have the inequality

$$G'(z, \zeta, f) \leq y(z, (f-\zeta)^{-1}),$$

which together with (65) proves relations (63) and (64). In particular this result implies that

$$G(z, \zeta, f) < \infty, \qquad f(z) \neq \zeta. \tag{66}$$

6D. Conversely, by (66) the characteristic of f remains bounded for all admissible Ω. As the first step in the proof of this assertion we show that for a fixed z the function $G(z, \zeta, f)$ is lower semicontinuous in ζ. To each ζ_0 and $\varepsilon > 0$ there correspond points a_i with $f(a_i) = \zeta_0$, $i = 1, \cdots, n$, such that

$$G(z, \zeta_0, f) \leq \sum_{i=1}^{n} \nu(f, a_i) g(z, a_i) + \varepsilon.$$

Since f is meromorphic there exists a positive constant A with the property that every ζ satisfying the relation $|\zeta - \zeta_0| < \eta < A$ is taken by f exactly $\nu(f, a_i)$ times in a neighborhood $N_i(\eta)$ of a_i. This neighborhood shrinks to a_i as $\eta \to 0$. Since $g(z, a)$ is continuous as a function of a, we deduce that

$$G(z, \zeta_0, f) \leq \liminf_{\zeta \to \zeta_0} G(z, \zeta, f) + \varepsilon.$$

This holds for every $\varepsilon > 0$, and our claim follows.

6E. If (66) holds we let z_0 be an arbitrary point of R and consider a closed disk D: $|\zeta - \zeta_0| \leq r$ such that $f(z_0) \notin D$. For any positive integer n, D_n is the set of points ζ in D such that $G(z_0, \zeta, f) \leq n$. The set D is the union of the sets D_n, each of which is closed because of the lower semicontinuity of G. Since D is of second category, it follows that for some n the set D_n contains an interior point and therefore some circle C: $|\zeta - \zeta_1| = \rho$. We conclude that if ζ is a point of the circle C, then $G(z_0, \zeta, f) \leq n$, or $G(z_0, e^{i\vartheta}, \rho^{-1}(f - \zeta_1)) \leq n$ with $0 \leq \vartheta < 2\pi$. For any regular region containing z_0 as an interior point we have the inequality

$$N(\Omega, e^{i\vartheta}, \rho^{-1}(f-\zeta_1)) \leq G(z_0, e^{i\vartheta}, \rho^{-1}(f-\zeta_1)),$$

where N is defined with respect to z_0. It follows from Cartan's formula (56) that

$$T(\Omega, \rho^{-1}(f-\zeta_1)) \leq \overset{+}{\log} |\rho^{-1}(f(z_0)-\zeta_1)| + n.$$

We infer that $\rho^{-1}(f-\zeta_1) \in MB$ and hence $f \in MB$.

In summary we state (Heins [3]):

Theorem. *A meromorphic function f defined on a hyperbolic Riemann surface R belongs to the class MB if and only if*

$$G(z, \zeta, f) < \infty \tag{67}$$

for all z, ζ *with* $f(z) \neq \zeta$.

This result again (cf. 2M) shows that the class MB does not depend on the choice of the point z_0. By using a suitable mass distribution in Frostman's formula (54) it can be shown that f is in MB if (66) holds for ζ in a set of positive capacity.

7. Decomposition by uniformization

7A. Let R be a hyperbolic Riemann surface and let (U, π) be the universal covering surface of R, π being the projection map. It is clear that U is hyperbolic, for if $g(z, a)$ is the Green's function on R, then $g(\pi(\zeta), a)$ is a positive superharmonic function on U. For this reason we can and will take U to be the unit disk.

We let $\mathscr{T}$ be the group of conformal self-transformations τ of U onto itself for which $\pi(\tau(\zeta)) = \pi(\zeta)$, $\zeta \in U$. If f is a single-valued mapping with domain R, then the function f^* defined by $f^*(\zeta) = f(\pi(\zeta))$ is automorphic relative to $\mathscr{T}$:

$$f^*(\tau(\zeta)) = f^*(\zeta). \tag{68}$$

Conversely if the function f^* defined on U satisfies equation (68), then the function f defined on R by

$$f(z) = f^*(\zeta), \qquad \pi(\zeta) = z$$

is single-valued with domain R. If one of the functions f or f^* is harmonic, superharmonic, analytic, or meromorphic, then the other has the same property.

We shall prove the following useful result due to P. Myrberg [3]:

Theorem. *Let* g_U *denote the Green's function on* U. *Then*

$$G(\zeta, z_0, \pi) = g(\pi(\zeta), z_0),$$

where

$$G(\zeta, z_0, \pi) = \sum_{\pi(\xi) = z_0} g_U(\zeta, \xi).$$

The function $g(\pi(\zeta), z_0)$ is a positive superharmonic function on U with logarithmic singularities at ξ, $\pi(\xi) = z_0$. Consequently if $\pi(\xi_1) = z_0$ we have the inequality

$$g(\pi(\zeta), z_0) \geq g_U(\zeta, \xi_1).$$

A fortiori $g(\pi(\zeta), z_0) - g_U(\zeta, \xi_1)$ is a positive superharmonic function on U with logarithmic singularities at the points $\xi \neq \xi_1$ which satisfy the relation $\pi(\xi) = z_0$. (Here and in the remainder of this section we include the constant 0 in the class of positive functions.) It follows that if $\pi(\xi_2) = z_0$ and $\xi_2 \neq \xi_1$, then

$$g(\pi(\zeta), z_0) - g_U(\zeta, \xi_1) \geq g_U(\zeta, \xi_2).$$

Repeating this argument we find that

$$g(\pi(\zeta), z_0) \geq G(\zeta, z_0, \pi). \tag{69}$$

On the other hand, $G(\zeta, z_0, \pi)$ is automorphic relative to $\mathscr{T}$ and for this reason defines a positive superharmonic function φ on R with a logarithmic singularity at z_0:

$$\varphi(z) = G(\zeta, z_0, \pi), \qquad \pi(\zeta) = z.$$

From this equality one infers that

$$G(\zeta, z_0, \pi) \geq g(\pi(\zeta), z_0),$$

which with inequality (69) completes the proof.

7B. It follows from Theorem 6E and Myrberg's theorem that a meromorphic function f on the hyperbolic Riemann surface R belongs to MB if and only if f^* belongs to MB on U. In view of the classical decomposition (Nevanlinna [22, p. 191]) for MB-functions in the unit disk U, the function f is in MB if and only if

$$\log |f^*(\zeta)| = G(\zeta, \infty, f^*) - G(\zeta, 0, f^*) + H_0^*(\zeta),$$

where H_0^* is the difference of two positive harmonic functions on U. The term on the left as well as the first two terms on the right are evidently automorphic relative to $\mathscr{T}$ and consequently so is H_0^*. If we set

$$H_0(z) = H_0^*(\zeta), \qquad \pi(\zeta) = z$$

and use Myrberg's theorem we see that *f is in MB if and only if*

$$\log |f(z)| = G(z, \infty, f) - G(z, 0, f) + H_0(z), \tag{70}$$

where H_0 is the difference of two positive harmonic functions on R. This is again the decomposition theorem 3B in the case MB.

For later reference we add here the following observation. Since $f - \xi$ and f are simultaneously in the class MB, f is in MB if and only if for some (and hence every) complex number ζ

$$\log |f(z) - \zeta| = G(z,\infty,f) - G(z,\zeta,f) + H_\zeta(z), \tag{71}$$

where H_ζ is the difference of two positive harmonic functions on R.

Remark. Parreau [2] gave another condition for a meromorphic function f on R to belong to MB: the function $\log^+ |f|$ has a harmonic majorant on $R - f^{-1}(\infty)$. The necessity of his criterion follows from the decomposition theorem, and the sufficiency is given by Corollary 3 in 4A. The function $u(z, f)$ of Theorem 3A is the least harmonic majorant of $\log^+ |f|$.

8. Theorems of Heins, Parreau, and Rao

8A. Let HB be the class of bounded harmonic functions. We are going to prove a result having as a consequence the inclusion $O_{HB} \subset O_{MB}$, first established by Heins [3]. We denote by $P(B)$ the set of nonnegative harmonic functions on R.

Definition. *A function h belonging to $P(R)$ is quasi-bounded if $h = \lim_{n \to \infty} h_n$, where $\{h_n\}$ is an increasing sequence of bounded members of $P(R)$. The function s is singular if the only bounded member of $P(R)$ majorized by s is the constant zero.*

If the function s is singular and h is quasi-bounded and $h \leq s$, then $h = 0$. We shall use the following well-known decomposition due to Parreau [2]:

Every function h in $P(R)$ has a unique representation

$$h = h_B + h_S, \tag{72}$$

where h_B is quasi-bounded and h_S is singular. Furthermore

$$h_B = \lim_{n \to \infty} \text{G.H.M.} \min(h, n).$$

Here G.H.M. stands for the "greatest harmonic minorant of". The functions h_B and h_S are called the quasi-bounded and singular parts of h.

We remark in passing that the above result can be deduced from the Poisson-Stieltjes representation for positive harmonic functions in the unit disk and the ideas used in Lemmas 8C and 8D.

8B. If f belongs to the class MB on R, then it follows from relation (71) that

$$\log |f(z) - \zeta| = G(z, \infty, f) - G(z, \zeta, f) + P_\zeta(z) - P'_\zeta(z), \tag{73}$$

where P_ζ and P'_ζ belong to $P(R)$. We shall establish the following result due to Rao [1], [2]:

Theorem. *If a nonconstant f is in MB on a Riemann surface R, then there exists at most one complex number ζ such that the difference between the quasi-bounded parts of P_ζ and P'_ζ is constant.*

The theorem has an immediate consequence:

Corollary. *If the only bounded harmonic functions on the Riemann surface R are constants, then every meromorphic function of bounded characteristic on R is also constant:*

$$O_{HB} \subset O_{MB}.$$

In fact, for a nonconstant $f \in MB$ any ζ not satisfying the condition of the theorem would give a nonconstant quasi-bounded function, hence a nonconstant bounded function.

To prove the theorem we need Lemmas 8C to 8F.

8C. We shall use the notations of 6 and 7.

Lemma. *A function s in $P(R)$ is singular if and only if s^* is a singular member of $P(U)$.*

We shall indicate a proof only for the necessity of the condition. The proof of the sufficiency can be carried out using similar arguments.

Suppose that s is singular. We must then show that the quasi-bounded part s_B^* of s^* vanishes. We set

$$v_n = \text{G.H.M.} \min (s^*, n),$$

and conclude by 8A that

$$s_B^* = \lim_{n \to \infty} v_n.$$

It will suffice to show that $v_n \equiv 0$ for all n.

It follows from the definitions of v_n and s^* that

$$v_n(\zeta) \leq s^*(\zeta) = s(\pi(\zeta)). \tag{74}$$

Consequently if τ is in $\mathcal{T}$, then

$$v_n(\tau(\zeta)) \leq s^*(\tau(\zeta)) = s(\pi(\tau(\zeta))) = s(\pi(\zeta)) = s^*(\zeta),$$

and a fortiori $v_n(\tau(\zeta)) \leq n$. Therefore

$$v_n(\tau(\zeta)) \leq \text{G.H.M.} \min (s^*(\zeta), n) = v_n(\zeta).$$

Since this holds for all points ζ in U and every τ belonging to the group $\mathcal{T}$, the reverse inequality is also valid. We infer that v_n is automorphic relative to $\mathcal{T}$ and that the equation

$$v_n'(z) = v_n(\zeta), \qquad \pi(\zeta) = z,$$

defines a bounded harmonic function v_n' which because of relation (74) satisfies the inequality

$$v_n'(z) \leq s(z).$$

The function s is singular, and it follows that v_n' and consequently v_n is zero.

8D. We next consider radial limits:

Lemma. *A singular function S in $P(U)$ has the radial limit zero for all but perhaps a set of points of measure zero.*

We let $\varphi(t)=\lim_{r\to 1} S(re^{it})$ be the radial limit of S which exists for almost all t (e.g., Nevanlinna [22, p. 201]). We set $\varphi_n(t)=\min(\varphi(t), n)$ and let $K(z, re^{it})$ be the Poisson kernel on U. If $|z|<r<1$, we find using Fatou's lemma that

$$S(z) = \frac{1}{2\pi}\int_0^{2\pi} S(re^{it})K(z, re^{it})\, dt$$

$$\geq \frac{1}{2\pi}\int_0^{2\pi} \liminf_{r\to 1} S(re^{it})K(z,e^{it})\, dt.$$

As a consequence we have the inequality

$$S(z) \geq \frac{1}{2\pi}\int_0^{2\pi} \varphi_n(t)K(z, e^{it})\, dt.$$

Since the right-hand side represents a bounded harmonic function defined on U and having the radial limits $\varphi_n(t)$, we conclude that $\varphi_n(t)=0$ almost everywhere. This proves the lemma, for $\varphi(t)=\lim_{n\to\infty}\varphi_n(t)$.

8E. For complete proofs of the well-known Lemmas 8E and 8F we refer to Nevanlinna [22, pp. 207–208].

Lemma. *A convergent sum of Green's functions on U has the radial limit zero almost everywhere.*

8F. We also have:

Lemma. *A function of bounded characteristic on U has radial limits almost everywhere and the set of its radial limits which correspond to a set on the boundary of positive Lebesgue measure contains more than two points.*

8G. We shall now prove Rao's theorem. Suppose that it is not true. Then for some function f of bounded characteristic on R and two complex numbers $\zeta_1 \neq \zeta_2$

$$\log|f(z)-\zeta_i| = G(z, \infty, f) - G(z, \zeta_i, f) + k_i + s_i(z) - s_i'(z), \qquad i = 1, 2,$$

where s_i and s_i' are singular members of $P(R)$ and k_i is a constant. From this relation we infer using Myrberg's theorem and Lemma 8C that for $i=1, 2$

$$\log|f^*(\zeta)-\zeta_i| = G(\zeta, \infty, f^*) - G(\zeta, \zeta_i, f^*) + k_i + S_i(\zeta) - S_i'(\zeta),$$

where S_i and S_i' are singular members of $P(U)$. Since the function f^* is of bounded characteristic on U, we deduce from Lemma 8F that f^* has radial limits $f^*(e^{i\vartheta})$ for almost all ϑ. Consequently we have the equation

$$\log |f^*(e^{i\vartheta}) - \zeta_i| = k_i, \qquad i = 1, 2,.$$

for almost all ϑ, as is seen by applying Lemmas 8D and 8E. In other words, the radial limits of f^* lie on both circles

$$|\zeta - \zeta_i| = e^{k_i}, \qquad i = 1, 2, \quad \zeta_1 \neq \zeta_2.$$

In view of the second part of Lemma 8F this is a contradiction, and the theorem is proved.

Remarks. It is possible that the difference between the quasi-bounded parts P_ζ and P'_ζ reduces to a constant for some ζ. For example, if f is the identity function on the Riemann surface $R = U$, then the above difference *vanishes* for $\zeta = 0$. An interesting problem is to characterize the surfaces R and functions f for which no exceptional ζ exists.

8H. On combining the inclusion relations we find that

$$O_G = O_{M_eB} \subset O_{HB} \subset O_{MB} \subset O_{AB}. \tag{75}$$

Consider a surface R with a finite number of sheets over the complex plane. Then the projection map has bounded characteristic if $R \notin O_G$. This follows immediately from Theorem 6E. Consequently the intersections of this class of surfaces R with O_{MB} and with O_G are identical. In particular on every hyperbolic surface of finite genus there is a nonconstant function of bounded characteristic. This is no longer true for hyperbolic surfaces of infinite genus. In fact, the inclusion $O_{HB} \subset O_{MB}$ is strict. There also exists a planar hyperbolic surface that carries no nonconstant AB functions. These and other problems regarding the role of O_{MB} in the classification of Riemann surfaces are dealt with by Heins [3] and Rao [1] to [3].

CHAPTER IV

FUNCTIONS ON PARABOLIC RIEMANN SURFACES

In I.10 to 12 we encountered examples of R_p-surfaces, i.e., surfaces possessing capacity functions with compact level lines. We shall now draw such surfaces under a more systematic study and show that every parabolic surface is of type R_p.

To this end we give in §1 the solution of the problem of constructing the Evans-Selberg potential on an arbitrary parabolic surface. As an application we obtain in §2 the following extension of the af Hällström-Nevanlinna-Kametani theorem on exceptional sets: every function meromorphic and with the Weierstrass property in a boundary neighborhood of a parabolic Riemann surface assumes every value infinitely often except perhaps for a countable union of compact sets of capacity zero.

By virtue of this important result the present chapter forms a natural unity with the rest of the book, in particular the theorems in I.19A, II.16A–B, and V.1A on exceptional sets. Moreover, the main existence theorem II.4C on principal functions is used in constructing the Evans-Selberg potential in IV.5C, and other aspects of R_p-surfaces are discussed in VI.19.

The results in the present chapter were obtained by Nakai [1].

§1. THE EVANS-SELBERG POTENTIAL

The Evans-Selberg potential $p(z, z_0)$ with pole at z_0 on an open Riemann surface R is the harmonic function on $R - z_0$ such that $p(z, z_0) - \log |z - z_0|$ has a harmonic extension to z_0 and $\lim p(z, z_0) = \infty$ as z tends to the ideal boundary of R. In this section we shall prove that the existence of the Evans-Selberg potential on R is assured if and only if R is parabolic. The necessity is almost trivial, and our main task consists in constructing the Evans-Selberg potential on a parabolic surface.

The construction was discussed initially by Evans [1] and Selberg [11], then by Noshiro [4] and from a more general point of view by Rudin [1], Ugaeri [1], Hong [1], and Inoue [1], among others. These constructions presupposed the realization of the ideal boundary of R as a relative boundary with respect to a larger surface and thus were restricted to the case of finite genus. Kuramochi [3] made the first attempt at surfaces of infinite genus by realizing the ideal boundary as the Martin boundary. However, the easiest way to avoid topological difficulties is to use, following Nakai [1], the Čech boundary which in a sense is the finest or largest realization of the ideal boundary.

1. The Čech compactification

1A. By a *continuous function* f on a topological space X we mean a continuous map of X into the extended real line $[-\infty, \infty]$. In case $f(X) \subset (-\infty, \infty)$ we say that f is a *finitely continuous function* on X. We denote by $C(X)$ the totality of continuous functions on X. For topological tools such as Tychonoff's theorem on topological products and Urysohn's theorem on the extension of continuous functions we refer the reader to any standard text in topology (e.g., Kelley [1]).

Let R be a locally compact Hausdorff space, e.g., an open Riemann surface.

Definition. *The Čech compactification of R is the compact Hausdorff space $\tilde{R}$ satisfying the following conditions:*

(a) *R is a dense subspace of $\tilde{R}$,*
(b) $C(\tilde{R}) \mid R = C(R)$.

The mapping $\tilde{f} \to \tilde{f} \mid R$ of $C(\tilde{R})$ into $C(R)$ is injective by (a) and surjective by (b). Hence for any f in $C(R)$ there exists a unique $\tilde{f}$ in $C(\tilde{R})$ with $\tilde{f} \mid R = f$; i.e., any continuous function f on R can be extended uniquely to a continuous function on $\tilde{R}$. In view of this we shall sometimes loosely say that f is continuous on $\tilde{R}$ instead of continuously extended to $\tilde{R}$.

Let $C_0(R)$ be the subfamily of $C(R)$ consisting of functions f with compact supports S_f in R. Since S_f is also compact in $\tilde{R}$, $\tilde{R} - S_f$ is open and $R \cap (\tilde{R} - S_f) = R - S_f$. Since f vanishes on $R - S_f$, by (a) $\tilde{f}$ vanishes on $\tilde{R} - S_f \supset \tilde{R} - R$, where $\tilde{f} \mid R = f$. For any point z in R there exists an f in $C_0(R)$ with $f(z) \neq 0$. We conclude that

$$\tilde{R} - R = \bigcap_{f \in C_0(R)} N_{\tilde{f}},$$

where $N_{\tilde{f}} = \{x \mid x \in \tilde{R}, \tilde{f}(x) = 0\}$ with $\tilde{f} \mid R = f$ and $\tilde{f} \in C(\tilde{R})$. Since $N_{\tilde{f}}$ is compact in $\tilde{R}$, $\tilde{R} - R$ is compact in $\tilde{R}$.

The compact set

$$\Gamma = \tilde{R} - R$$

will be referred to as the *Čech boundary* of R.

1B. Next we observe that *the Čech compactification $\tilde{R}$ of R is unique up to a homeomorphism fixing R elementwise.*

In fact, let $\tilde{R}$ and R' be two Čech compactifications of R. We denote by I the identity map $I(z)=z$ of R onto itself. Let x be a point in $\tilde{R}-R$ and let $\{z_a\}$ be an arbitrary directed net in R converging to x. Assume that $\{z_a\}$ has two distinct accumulation points x_1' and x_2' in $R'-R$. Then there exists a function f' in $C(R')$ such that $f'(x_i')=i$ for $i=1, 2$. Let $f=f' \mid R$ and $\tilde{f}$ be in $C(\tilde{R})$ with $\tilde{f} \mid R=f$. Then

$$\lim_a f(z_a) = \lim_a \tilde{f}(z_a) = \tilde{f}(x),$$

while

$$\limsup_a f(z_a) = \limsup_a f'(z_a) \geq f'(x_2') = 2$$

and

$$\liminf_a f(z_a) = \liminf_a f'(z_a) \leq f'(x_1') = 1.$$

This is a contradiction, and we conclude that the map $I: R \to R$ is extended to a continuous map $\tilde{I}$: $\tilde{R} \to R'$ and similarly to a continuous map I': $R' \to \tilde{R}$. But since $I' \circ \tilde{I}$ and $\tilde{I} \circ I'$ are continuous maps of $\tilde{R}$ and R' into themselves fixing R elementwise, we see by (a) that these are identity maps of $\tilde{R}$ and R'. Thus $\tilde{I}=I'^{-1}$ is a homeomorphism of $\tilde{R}$ onto R' fixing R elementwise.

1C. Finally we shall establish *the existence of the Čech compactification* $\tilde{R}$ *of* R.

For each f in $C(R)$ we denote by I_f the closure $\overline{f(R)}$ of the range set $f(R)$ in $[-\infty, \infty]$. Since $[-\infty, \infty]$ is compact, I_f is also a compact Hausdorff space. Consider the topological product of $I_f, f \in C(R)$, with the weak topology:

$$\Delta = \prod_{f \in C(R)} I_f.$$

By Tychonoff's theorem Δ is again a compact Hausdorff space. We denote by π_f the projection map of Δ onto I_f. Let φ be the map of R into Δ defined by

$$\varphi(z) = \prod_{f \in C(R)} f(z),$$

i.e., $\pi_f(\varphi(z))=f(z)$. It is easy to see by the definition of the weak topology that φ is a homeomorphism of R onto $\varphi(R)$. Hence it is sufficient to demonstrate the existence of the Čech compactification of $\varphi(R)$.

We shall show that the Čech compactification of $\varphi(R)$ is the closure $\overline{\varphi(R)}$ of $\varphi(R)$ in Δ. Since $\overline{\varphi(R)}$ is a closed subset of a compact Hausdorff space Δ, $\overline{\varphi(R)}$ itself is a compact Hausdorff space. Condition (a) is clearly satisfied. To prove that $\overline{\varphi(R)}$ satisfies (b) it is sufficient to show that for any F in $C(\varphi(R))$ there exists an $\bar{F}$ in $C(\overline{\varphi(R)})$ such that $\bar{F}\mid\varphi(R)=F$. Obviously $F\circ\varphi=f\in C(R)$ and therefore F has the representation $F(\varphi(z))=f(z)=\pi_f(\varphi(z))$. For any $x\in\Delta$ define the function $\bar{F}$ by

$$\bar{F}(x) = \pi_f(x).$$

Since π_f is the projection of Δ onto I_f, it is clearly continuous on Δ and hence on $\overline{\varphi(R)}$. Thus $\bar{F}\in C(\overline{\varphi(R)})$. But on $\varphi(R)$, $\bar{F}(\varphi(z))=\pi_f(\varphi(z))=F(\varphi(z))$, i.e., $\bar{F}\mid\varphi(R)=F$.

2. Green's kernel on the Čech compactification

2A. Let Ω be a regular region on a parabolic Riemann surface R.

Lemma. *Let s be a superharmonic function on $R-\bar{\Omega}$ bounded from below and continuous on $R-\Omega$. Then s is continuous on $\tilde{R}-\Omega$ and $s\geq\min_{\partial\Omega}s$ on $\tilde{R}-\Omega$.*

Proof. We can extend s continuously to all of R. The function s thus extended is continuous on $\tilde{R}$ and consequently s is continuous on $\tilde{R}-\Omega$. Assume that $s\geq c>-\infty$ on $R-\bar{\Omega}$. Let $\{R_n\}_1^\infty$ be an exhaustion of R, i.e., a sequence of regular regions of R with $\bar{R}_n\subset R_{n+1}$ and $\bigcup R_n=R$. Suppose that $\bar{\Omega}\subset R_1$ and let u_n be harmonic on $\bar{R}_n-\Omega$ with $u_n\mid\partial R_n=c$ and $u_n\mid\partial\Omega=m=\min_{\partial\Omega}s$. Then by the maximum principle $s\geq u_n$ on $\bar{R}_n-\Omega$. But since R is parabolic, $u_n\to m$ as $n\to\infty$. Thus $s\geq m$ on $R-\Omega$ and hence on $\tilde{R}-\Omega$ (cf. App.I.4).

2B. Let R_0 be a regular region with connected complement $R-\bar{R}_0$ and let $g_a(z)$ be the Green's function on $R-\bar{R}_0$ with its pole at $a\in R-\bar{R}_0$. We set $g_a(z)=0$ if one or both of z and a belong to ∂R_0. It is bounded except in an arbitrary neighborhood of a. In view of Lemma 2A it is uniquely determined. For the existence of the Green's function and its properties we refer the reader to Ahlfors-Sario [1, pp. 188–189]. In particular, the function is symmetric:

$$g_a(z) = g_z(a). \tag{1}$$

2C. We next show that *the function $(z,a)\to g_a(z)$ is continuous on $(R-R_0)\times(R-R_0)-\partial R_0\times\partial R_0$.* To this end we first prove:

Lemma. *Let D be a plane region and X a topological space. If $h(z,x)$ is a real-valued function bounded from below such that $z\to h(z,x)$ is harmonic in D and $x\to h(z,x)$ is continuous on X, then $(z,x)\to h(z,x)$ is continuous on $D\times X$.*

Proof. We may assume that $h>0$. Fix a point (z_0, x_0) in $D\times X$. Choose r so that the closed disk $\{z \mid |z-z_0|\le r\}\subset D$. Then by Poisson's formula

$$\frac{r-\rho}{r+\rho}h(z_0, x) \le h(z, x) \le \frac{r+\rho}{r-\rho}h(z_0, x), \qquad \rho = |z-z_0|$$

and consequently $\lim_{(z,x)\to(z_0,x_0)}h(z, x)=h(z_0, x_0)$.

From this lemma we conclude that $(z, a)\to g_a(z)$ is continuous at (z_0, a_0) with $a_0\ne z_0$. Next assume that $a_0=z_0\in R-\bar{R}_0$. Consider the closed disks D: $|z-z_0|\le 1/2$ and D': $|z-z_0|\le 1/4$. Since $g_a(z)$ is continuous on $\partial D\times D'$, there exists a positive constant M such that $g_a(z)\ge M$ on $\partial D\times D'$. For $(z, a)\in D\times D'$ we have

$$g_a(z) = \log\frac{1}{|z-a|} + h(z, a),$$

where $h(z, a)$ is harmonic with respect to z. Clearly $-\log|z-a|\le\log 4$ on $\partial D\times D'$ and therefore

$$h(z, a) \ge M-\log 4$$

on $\partial D\times D'$. By the harmonicity of $z\to h(z, a)$ in D the same is true on $D\times D'$, and we have

$$g_a(z) \ge \log\frac{1}{|z-a|} + M-\log 4$$

for all $(z, a)\in D'\times D'$. We conclude that $\lim_{(z,a)\to(z_0,z_0)}g_a(z)=\infty=g_{z_0}(z_0)$.

2D. By Lemma 2A, $a\to g_a(z)$ is continuous on $\tilde{R}-R_0$. We denote by $\tilde{g}_a(z)$ the extended function of a. Once we fix a in $\tilde{R}-\bar{R}_0$, $z\to\tilde{g}_a(z)$ is again a real-valued function on $R-\bar{R}_0$. We shall show:

Lemma. *The function $z\to\tilde{g}_a(z)$ is a strictly positive harmonic function on $R-\bar{R}_0$, except perhaps at a, with continuously vanishing boundary values on ∂R_0.*

Proof. Since $\tilde{g}_a(z)=g_a(z)$ if $a\in R-\bar{R}_0$, we have only to consider the case $a=\tilde{a}\in\Gamma$. In view of Lemma 2A, $\tilde{g}_{\tilde{a}}(z)>0$ for any $z\in R-\bar{R}_0$. Fix a point $z_0\in R-\bar{R}_0$ and take the disks D: $|z-z_0|\le 2$ and D': $|z-z_0|\le 1/4$ and a subsurface F of R, with $R-\bar{F}$ a regular region such that $\bar{R}_0\cup D\subset R-\bar{F}$. Then $g_a(z)$ is finitely continuous for (a, z) in $\partial F\times D'$. By the harmonicity and boundedness of $a\to g_a(z)$, $z\in D'$, on F and by Lemma 2A we conclude that there exists a positive constant M such that $g_a(z)<M$ for (a, z) in $F\times D'$. By using the Poisson representation as in the proof of Lemma 2C we see that

$$|g_a(z)-g_a(z')| \le g_a(z)\rho(|z-z'|) \le M\rho(|z-z'|),$$

where

$$\rho(|z-z'|) = \max\left(1-\frac{1-|z-z'|}{1+|z-z'|}, \frac{1+|z-z'|}{1-|z-z'|}-1\right).$$

Thus the family $\{g_a(z) \mid a \in F\}$ of functions $z \to g_a(z)$ on D' is compact with respect to the parameter set F and $g_a(z)$ converges uniformly to $\tilde{g}_{\tilde{a}}(z)$ on D' as $a \to \tilde{a}$, $a \in F$. We conclude that $\tilde{g}_{\tilde{a}}(z)$ is harmonic in D' and consequently in $R-\bar{R}_0$.

Next take a point z_0 on ∂R_0 and let D and D' be as above such that $\partial R_0 \cap D = \{z \mid |z-z_0| \le 2, \operatorname{Im}(z-z_0)=0\}$. By the preceding argument there exists a constant N such that $g_a(z) < N$ for $(a, z) \in F \times (D \cap (R-R_0))$. Let $u_a(z) = g_a(z)+N$ for $(a, z) \in F \times (D \cap (R-R_0))$ and $u_a(z) = -g_a(\bar{z})+N$ for $(a, z) \in F \times (D \cap R_0)$. Then $z \to u_a(z)$ is harmonic in D and $0 < u_a(z) < 2N$ for $(a, z) \in F \times D$. By an argument similar to the one above we see that $u_a(z)$ converges uniformly on D' as $a \in F$ tends to $\tilde{a}$. Thus in particular $g_a(z) \to \tilde{g}_{\tilde{a}}(z)$ uniformly on $D' \cap (R-R_0)$ and we see that $\tilde{g}_{\tilde{a}}$ vanishes on ∂R_0.

2E. In view of Lemma 2D the function $z \to \tilde{g}_a(z) = \lim_{b\to a} g_b(z)$ with $b \in R-\bar{R}_0$ is nonnegative harmonic on $R-R_0$ for each a in $\tilde{R}-R_0$ and hence it can be extended continuously to $\tilde{R}-R_0$.

Definition. *The Green's kernel $G(x, y)$ on $\tilde{R}-R_0$ is the function on $(\tilde{R}-R_0)\times(\tilde{R}-R_0)$ defined by the double limit*

$$G(x, y) = \lim_{a\to x}\left(\lim_{b\to y} g_b(a)\right)$$

with $a, b \in R-\bar{R}_0$.

Clearly if we fix y in $\tilde{R}-R_0$, then $G(z, y) = \tilde{g}_y(z)$ on $R-R_0$. Thus $x \to G(x, y)$ is the continuous extension of the function $z \to \tilde{g}_y(z)$ to $\tilde{R}-R_0$.

2F. We shall show:

Lemma. *The Green's kernel $G(x, y)$ on $\tilde{R}-R_0$ has the following properties:*

(a) *$G(z, t) = g_t(z)$ for (z, t) in $(R-R_0)\times(R-R_0)$,*

(b) *$G(z, x) = G(x, z)$ for z in $R-R_0$ and x in $\tilde{R}-R_0$,*

(c) *$G(x, y)$ is continuous in x on $\tilde{R}-R_0$ for fixed y in $\tilde{R}-R_0$,*

(d) *$G(z, x)$ is continuous in (z, x) on $(R-R_0)\times(\tilde{R}-R_0)-\partial R_0\times\partial R_0$ and finitely continuous in (z, x) on $(R-R_0)\times\Gamma$,*

(e) *$G(z, x)$ is harmonic in z on $R-R_0-x$ for fixed x in $\tilde{R}-R_0$,*

(f) *$G(x, y) > 0$ (resp. $\equiv 0$) on $\tilde{R}-\bar{R}_0$ for any fixed y in $\tilde{R}-\bar{R}_0$ (resp. ∂R_0),*

(g) *$\int_{\partial R_0} dG^*(z, x) = 2\pi$ for any fixed x in $\tilde{R}-\bar{R}_0$, where ∂R_0 is oriented positively with respect to R_0,*

(h) *$G(z, x) = 0$ on ∂R_0 for any fixed x in $\tilde{R}-R_0$.*

Proof. Properties (a) and (c) are obvious by Definition 2E. Property (b) follows from Definition 2E and (1), i.e.,

$$\begin{aligned} G(z, x) &= \lim_{a \to z} \left(\lim_{b \to x} g_b(a) \right) \\ &= \lim_{a \to z} \tilde{g}_x(a) = \tilde{g}_x(z) = \lim_{b \to x} g_b(z) \\ &= \lim_{b \to x} \left(\lim_{a \to z} g_b(a) \right) \\ &= \lim_{b \to x} \left(\lim_{a \to z} g_a(b) \right) \\ &= G(x, z), \end{aligned}$$

where $a, b \in R - \bar{R}_0$. Properties (e), (f), and (h) are obvious in view of Lemma 2D.

Next we prove (d). The continuity of $G(z, x)$ at $(z_0, x_0) \in (R - R_0) \times (R - R_0) - \partial R_0 \times \partial R_0$ follows from 2C and (a). To prove the continuity at (z_0, x_0) in $(R - R_0) \times \Gamma$ let $D' = \{z \mid |z - z_0| \leq 1/4\}$ and R_1 be a regular region such that $\bar{R}_0 \cup D' \subset R_1$. In case $z_0 \in \partial R_0$ we moreover assume that $\partial R_0 \cap D' = \{z \mid |z - z_0| \leq 1/4, \operatorname{Im}(z - z_0) = 0\}$ and we extend the harmonic function $z \to G(z, x)$, $x \in \tilde{R} - R_1$, to D' harmonically; this is possible by (h). Since $G(z, x) = \tilde{g}_x(z)$ we conclude in the same manner as in 2D that $G(z, x)$ is bounded for $(z, x) \in D' \times (\tilde{R} - R_1)$. By (e), (c), (b), and Lemma 2C, $G(z, x)$ is finitely continuous at $(z_0, x_0) \in (R - R_0) \times \Gamma$.

Finally we establish (g). First assume that $a \in R - \bar{R}_0$. Let $\{R_n\}_1^\infty$ be an exhaustion of R with $\bar{R}_0 \subset R_1$, $a \in R_1$, and D_a: $|z - a| < 1$ with $\bar{D}_a \subset R_1 - \bar{R}_0$. Denote by v_n the harmonic function in $\bar{R}_n - R_0 - D_a$ such that $v_n \mid \partial R_0 \cup \partial D_a = 1$ and $v_n \mid \partial R_n = 0$. Since R is parabolic, $v_n \to 1$ on $R - \bar{R}_0 - \bar{D}_a$ and $D_{R_n - \bar{R}_0 - \bar{D}_a}(v_n) \to 0$ as $n \to \infty$. By Green's formula

$$\begin{aligned} D_{R_n - \bar{R}_0 - \bar{D}_a}(v_n, g_a) &= \int_{\partial(R_n - \bar{R}_0 - \bar{D}_a)} v_n \, dg_a^* \\ &= - \int_{\partial R_0} dg_a^* - \int_{\partial D_a} dg_a^*, \end{aligned}$$

where ∂R_0 and ∂D_a are oriented positively with respect to R_0 and D_a. Since $g_a(z) = -\log |z - a| + h(z)$ on D_a with harmonic h,

$$\int_{\partial D_a} dg_a^* = -2\pi,$$

while

$$|D_{R_n - \bar{R}_0 - \bar{D}_a}(v_n, g_a)|^2 \leq D_{R_n - \bar{R}_0 - \bar{D}_a}(g_a) \cdot D_{R_n - \bar{R}_0 - \bar{D}_a}(v_n) \to 0$$

as $n \to \infty$. Thus $\int_{\partial R_0} dg_a^* = 2\pi$, i.e.,

$$\int_{\partial R_0} dG^*(z, a) = 2\pi \tag{2}$$

for a in $R - \bar{R}_0$.

Next let $x \in \Gamma$. For $a \to x$, $G(z, a)$ converges uniformly to $G(z, x)$ on $\bar{R}_1 - R_0$ as in 2D. On letting $a \to x$ in (2) with $a \in R - \bar{R}_0$ we obtain (g).

3. Transfinite diameter

3A. Again let R be a parabolic Riemann surface and $\tilde{R}$ (resp. Γ) the Čech compactification (resp. boundary) of R.

Given a nonempty compact set K in $\tilde{R} - \bar{R}_0$ we define $D_n(K)$ by

$$\binom{n}{2} D_n(K) = \inf_{x_1, \cdots, x_n \in K} \sum_{i<j}^{1, \cdots, n} G(x_i, x_j).$$

For arbitrary points $x_1, \cdots, x_{n+1}$ in K

$$\sum_{i<j}^{1, \cdots, n+1} G(x_i, x_j) = \sum_{i=1}^{k-1} G(x_i, x_k) + \sum_{j=k+1}^{n+1} G(x_k, x_j) + \sum_{i<j;\, i,j \neq k}^{1, \cdots, n+1} G(x_i, x_j).$$

Hence

$$\sum_{i<j}^{1, \cdots, n+1} G(x_i, x_j) \geq \sum_{i=1}^{k-1} G(x_i, x_k) + \sum_{j=k+1}^{n+1} G(x_k, x_j) + \binom{n}{2} D_n(K).$$

On summing these $n+1$ inequalities for $k = 1, \cdots, n+1$ we obtain

$$(n+1) \sum_{i<j}^{1, \cdots, n+1} G(x_i, x_j) \geq 2 \sum_{i<j}^{1, \cdots, n+1} G(x_i, x_j) + (n+1) \binom{n}{2} D_n(K),$$

that is,

$$(n-1) \sum_{i<j}^{1, \cdots, n+1} G(x_i, x_j) \geq (n+1) \binom{n}{2} D_n(K).$$

We conclude that $(n-1)\binom{n+1}{2} D_{n+1}(K) \geq (n+1)\binom{n}{2} D_n(K)$ and consequently

$$D_{n+1}(K) \geq D_n(K).$$

Hence we can define

$$D(K) = \lim_{n \to \infty} D_n(K), \tag{3}$$

which we shall call the *transfinite diameter* of K. It is easy to see that

$$D_n(K) \geq D_n(K') \quad \textit{and} \quad D(K) \geq D(K') \quad \textit{for} \quad K \subset K'. \tag{4}$$

3B. Given a nonempty compact set K in $\tilde{R}-\bar{R}_0$ we define $E_n(K)$ by

$$nE_n(K) = \sup_{x_1,\cdots,x_n \in K} \left(\inf_{x \in K} \sum_1^n G(x, x_i) \right).$$

For arbitrary points $x_1, \cdots, x_{n+m}$ in K

$$\inf_{x \in K} \sum_1^{n+m} G(x, x_i) \geq \inf_{x \in K} \sum_1^n G(x, x_i) + \inf_{x \in K} \sum_{n+1}^{n+m} G(x, x_i)$$

and hence

$$(5) \qquad (n+m)E_{n+m}(K) \geq nE_n(K)+mE_m(K).$$

Let $\alpha = \sup_n E_n(K)$. For any $\beta < \alpha$ we can find an m such that $E_m(K) > \beta$. Any positive integer n has a unique representation $n = qm+r$, where q and r are nonnegative integers such that $0 \leq r < m$. Then by (5), $E_{qm}(K) \geq E_m(K)$ for a positive integer q and again by (5)

$$nE_n(K) = (qm+r)E_{qm+r}(K) \geq qmE_{qm}(K)+rE_r(K) \geq qmE_m(K)$$

and therefore

$$E_n(K) \geq \frac{qm}{qm+r} E_m(K) > \frac{qm}{qm+r} \beta.$$

Here $q \to \infty$ as $n \to \infty$. Thus

$$\alpha \geq \limsup_{n \to \infty} E_n(K) \geq \liminf_{n \to \infty} E_n(K) \geq \beta.$$

On letting $\beta \to \alpha$ we conclude that

$$(6) \qquad E(K) = \lim_{n \to \infty} E_n(K)$$

exists. We shall call $E(K)$ *the (modified) Tchebycheff constant of* K.

3C. The following is one of the most important relations in our construction:

Lemma. *The transfinite diameter is dominated by the Tchebycheff constant,*

$$D(K) \leq E(K).$$

Proof. Let $n > 1$. We first show that there exists a system of n points $x_1, \cdots, x_n$ in K satisfying the equalities

$$(7) \qquad \sum_{j=n-i+1}^{n} G(x_{n-i}, x_j) = \inf_{x \in K} \sum_{j=n-i+1}^{n} G(x, x_j), \qquad i = 1, \cdots, n-1.$$

We can construct such a system inductively as follows. First choose x_n arbitrarily in K. Assume that $x_n, \cdots, x_{n-k+1}$ with $k \leq n-1$ have been already chosen so as to satisfy (7) for $i = 1, \cdots, k-1$. We have to select x_{n-k} satisfying (7) with $i = k$.

Consider the function

$$f(x) = \sum_{j=n-k+1}^{n} G(x, x_j).$$

It is continuous on $\tilde{R}-R_0$ by (c) of Lemma 2F. Hence we can find a point x_{n-k} on K such that

$$f(x_{n-k}) = \min_{x \in K} f(x),$$

which is (7) for $i=k$. From (7) and the definition of $E_i(K)$ it follows that

$$\sum_{j=n-i+1}^{n} G(x_{n-i}, x_j) \le iE_i(K), \qquad i = 1, \cdots, n-1.$$

On summing these $n-1$ inequalities we obtain

$$\sum_{i<j}^{1,\cdots,n} G(x_i, x_j) \le \sum_{1}^{n-1} iE_i(K).$$

This and the definition of $D_n(K)$ give

$$(8) \qquad D_n(K) \le \binom{n}{2}^{-1} \sum_{1}^{n-1} iE_i(K).$$

To prove $E(K) \ge D(K)$ we may assume that $E(K) < \infty$. Then by (5), $E_i(K) < \infty$. Let $a_i = |E(K)-E_i(K)|$ and $M_k = \sup_{i>k} a_i$. We have $M_n \to 0$ as $n \to \infty$ by (6), and consequently

$$\left|E(K) - \binom{n}{2}^{-1} \sum_{1}^{n-1} iE_i(K)\right| = \binom{n}{2}^{-1} \left|\sum_{1}^{n-1} i(E(K)-E_i(K))\right|$$

$$\le \binom{n}{2}^{-1} \sum_{1}^{n-1} ia_i = \binom{n}{2}^{-1} \sum_{1}^{k} ia_i + \binom{n}{2}^{-1} \sum_{k+1}^{n} ia_i$$

$$\le \binom{n}{2}^{-1} \binom{k}{2} M_1 + \binom{n}{2}^{-1} 2^{-1}(n-k)(n+k+1)M_k.$$

Here on first letting $n \to \infty$ and then $k \to \infty$ we obtain

$$E(K) = \lim_{n \to \infty} \binom{n}{2}^{-1} \sum_{1}^{n-1} iE_i(K).$$

For $n \to \infty$ in (8) we conclude that $D(K) \le E(K)$.

3D. Let Ω be a regular region in R with $\Omega \supset \bar{R}_0$.

Lemma. *The transfinite diameter of the Čech boundary of a parabolic Riemann surface satisfies the inequality*

$$D(\Gamma) \ge D(\tilde{R}-\Omega) = D(\partial\Omega).$$

Proof. The relations $D(\Gamma)\geq D(\tilde{R}-\Omega)$ and $D(\tilde{R}-\Omega)\leq D(\partial\Omega)$ follow easily from (4). Hence we have only to show that $D(\tilde{R}-\Omega)\geq D(\partial\Omega)$. To this end it is sufficient to prove that $D_n(\tilde{R}-\Omega)\geq D_n(\partial\Omega)$.

Let $x_1, \cdots, x_n$ be arbitrary points in $\tilde{R}-\Omega$. We choose n points $z_1, \cdots, z_n$ in $\partial\Omega$ satisfying the inequalities

$$(9)\quad \sum_{i<j}^{1,\cdots,k} G(z_i, z_j)+\sum_{i=1}^{k}\sum_{j=k+1}^{n} G(z_i, x_j)+\sum_{i<j}^{k+1,\cdots,n} G(x_i, x_j)$$
$$\leq \sum_{i<j}^{1,\cdots,k-1} G(z_i, z_j)+\sum_{i=1}^{k-1}\sum_{j=k}^{n} G(z_i, x_j)+\sum_{i<j}^{k,\cdots,n} G(x_i, x_j)$$

for $k=1, \cdots, n$. We can choose such a system inductively as follows.

Assume that $z_1, \cdots, z_{m-1}$, $m\leq n$, have been already chosen so as to satisfy (9) for $k=1, \cdots, m-1$. Consider the function

$$f(x) = \sum_{i=1}^{m-1} G(z_i, x)+\sum_{j=m+1}^{n} G(x, x_j).$$

By (b) of Lemma 2F

$$f(x) = \sum_{i=1}^{m-1} G(x, z_i)+\sum_{j=m+1}^{n} G(x, x_j).$$

By (e), (f), and (a) of Lemma 2F, $f(x)$ is positive continuous and superharmonic on $R-\Omega$ and it is continuous on $\tilde{R}-\Omega$. Hence by Lemma 2A there exists a point z_m in $\partial\Omega$ such that

$$f(z_m) = \min_{x\in\partial\Omega} f(x) \leq f(x)$$

for any x in $\tilde{R}-\Omega$. In particular $f(z_m)\leq f(x_m)$, i.e.,

$$\sum_{i=1}^{m-1} G(z_i, z_m)+\sum_{j=m+1}^{n} G(z_m, x_j) \leq \sum_{i=1}^{m-1} G(z_i, x_m)+\sum_{j=m+1}^{n} G(x_m, x_j).$$

It follows that

$$\sum_{i<j}^{1,\cdots,m} G(z_i, z_j)+\sum_{i=1}^{m}\sum_{j=m+1}^{n} G(z_i, x_j)+\sum_{i<j}^{m+1,\cdots,n} G(x_i, x_j)$$
$$= \sum_{i<j}^{1,\cdots,m-1} G(z_i, z_j)+\sum_{i=1}^{m-1}\sum_{j=m+1}^{n} G(z_i, x_j)+\sum_{i<j}^{m+1,\cdots,n} G(x_i, x_j)+f(z_m)$$
$$\leq \sum_{i<j}^{1,\cdots,m-1} G(z_i, z_j)+\sum_{i=1}^{m-1}\sum_{j=m+1}^{n} G(z_i, x_j)+\sum_{i<j}^{m+1,\cdots,n} G(x_i, x_j)+f(x_m)$$
$$= \sum_{i<j}^{1,\cdots,m-1} G(z_i, z_j)+\sum_{i=1}^{m-1}\sum_{j=m}^{n} G(z_i, x_j)+\sum_{i<j}^{m,\cdots,n} G(x_i, x_j).$$

This is (9) with $k=m$.

Thus we can construct a system $z_1, \cdots, z_n$ in $\partial\Omega$ satisfying (9) with $k=1, \cdots, n$. On summing these n inequalities we obtain

$$\sum_{i<j}^{1,\cdots,n} G(z_i, z_j) \leq \sum_{i<j}^{1,\cdots,n} G(x_i, x_j)$$

and consequently

$$\binom{n}{2} D_n(\partial\Omega) \leq \sum_{i<j}^{1,\cdots,n} G(x_i, x_j).$$

Since x $, \cdots, x_n$ are arbitrary in $\tilde{R}-\Omega$ we conclude that

$$\binom{n}{2} D_n(\partial\Omega) \leq \binom{n}{2} D_n(\tilde{R}-\Omega).$$

4. Energy integral

4A. Let R, $\tilde{R}$, Γ, and R_0 be as before. By a measure μ on $\tilde{R}-\bar{R}_0$ we mean a nonnegative regular Borel measure on $\tilde{R}-\bar{R}_0$ and we denote by S_μ the support of μ so that $\mu(\tilde{R}-\bar{R}_0-S_\mu)=0$. We only consider measures μ for which S_μ is compact in $\tilde{R}-\bar{R}_0$. The *Green's potential* $G_\mu(z)$ with respect to μ is defined by

$$G_\mu(z) = \int G(z, x)\, d\mu(x), \qquad z \in R-R_0.$$

It is easy to see that $G_\mu(z)$ is a positive superharmonic function on $R-\bar{R}_0$, harmonic on $R-\bar{R}_0-S_\mu$, and continuously zero on ∂R_0, if $\mu \neq 0$.

A subset A of $R-\bar{R}_0$ is said to be *polar* if there exists a positive superharmonic function on $R-\bar{R}_0$ which is infinite on A. A property is said to hold *quasi-everywhere* if it holds everywhere except on a polar set.

The following important fact, usually called the fundamental existence theorem in potential theory, was established by Frostman [3]:

Lemma. *Let s be a positive superharmonic function on $R-\bar{R}_0$ and K a compact set in $R-\bar{R}_0$. Then there exists a unique measure μ with $S_\mu \subset K$ such that*

(a) *$G_\mu \leq s$ on $R-\bar{R}_0$,*

(b) *$G_\mu = s$ quasi-everywhere on K.*

For the proof we refer the reader to Constantinescu-Cornea [2, pp. 40–41].

4B. For two measures μ and ν with compact supports in $R-\bar{R}_0$ we denote by $\|\mu\|^2$ the *energy* of μ and by $\langle\mu, \nu\rangle$ the *mutual energy* of μ and ν, defined by

$$\|\mu\|^2 = \int\int G(z, t)\, d\mu(z)\, d\mu(t),$$

$$\langle\mu, \nu\rangle = \int\int G(z, t)\, d\mu(z)\, d\nu(t).$$

Clearly $\langle\mu, \nu\rangle = \langle\nu, \mu\rangle$ and $\langle\mu, \mu\rangle = \|\mu\|^2$. Moreover we have the so-called energy principle:

Lemma. *The mutual energy of μ and ν is related to the energies of μ and ν by the inequality*

$$\langle\mu, \nu\rangle \leq ||\mu|| \cdot ||\nu||.$$

For the proof we again refer to Constantinescu-Cornea [2, pp. 46–47].

4C. For a nonempty compact set K in $R-\bar{R}_0$ we denote by m_K the totality of unit measures μ with $S_\mu \subset K$ and set

$$W(K) = \inf_{\mu \in m_K} \|\mu\|^2.$$

The quantity $1/W(K)$ is called the *capacity* of K.

Lemma. *Let Ω be a regular region of R with $\bar{R}_0 \subset \Omega$ and let u_Ω be the harmonic function on $\bar{\Omega}-R_0$ with $u_\Omega \mid \partial R_0 = 0$ and $u_\Omega \mid \partial\Omega = 1$. Then*

$$W(\partial\Omega) = \frac{2\pi}{D_{\Omega-\bar{R}_0}(u_\Omega)}.$$

For illustration, if $R_0 : |z| < 1$ and $\Omega : |z| < e$, then $W(\partial\Omega) = 1$.

Proof. Set $u_\Omega = 1$ on $R-\Omega$. Then u_Ω is positive superharmonic in $R-\bar{R}_0$ and, using Lemma 4A, we can find a unique measure μ with $S_\mu \subset \partial\Omega$ such that $G_\mu \leq u_\Omega \leq 1$ on $R-\bar{R}_0$ and $G_\mu = u_\Omega = 1$ quasi-everywhere on $\partial\Omega$. Let s be positive superharmonic in $R-\bar{R}_0$ and infinite on the subset of $\partial\Omega$ on which $G_\mu \neq 1$. Then we have

$$u_\Omega \leq G_\mu + \frac{s}{n}$$

on $\partial R_0 \cup \partial\Omega$ for any $n = 1, 2, \cdots$. By the same argument as in 2A, $u_\Omega \leq G_\mu + sn^{-1}$ on $R-\bar{\Omega}$ and clearly the same is true on $\Omega-\bar{R}_0$ and a fortiori on $R-R_0$.

We let $n \to \infty$ and see that $u_\Omega \leq G_\mu$ on $R-\bar{R}_0$ except perhaps on a polar set and, since $G_\mu \leq u_\Omega$,

$$u_\Omega = G_\mu$$

on $R-\bar{R}_0$ except perhaps on a polar set. But since such a set is of

Lebesgue measure zero, we must have $u_\Omega = G_\mu$ on $R - R_0$. Thus in particular $\mu(\partial\Omega) > 0$. For $\nu \in m_{\partial\Omega}$ we have by Lemma 4B

$$1 = \left(\int G_\mu \, d\nu\right)^2 = \langle \mu, \nu \rangle^2 \le \|\mu\|^2 \|\nu\|^2$$

and

$$\|\nu\|^2 \ge \frac{1}{\|\mu\|^2}.$$

On the other hand,

$$\mu(\partial\Omega) = \int d\mu = \int G_\mu \, d\mu = \|\mu\|^2.$$

If we set $\mu_\Omega = \mu/\mu(\partial\Omega)$, then $\mu_\Omega \in m_{\partial\Omega}$ and

$$\|\mu_\Omega\|^2 = \frac{\|\mu\|^2}{\mu(\partial\Omega)^2} = \frac{\|\mu\|^2}{\|\mu\|^4} = \frac{1}{\|\mu\|^2} \le \|\nu\|^2.$$

Thus

$$\|\mu_\Omega\|^2 = \inf_{\nu \in m_{\partial\Omega}} \|\nu\|^2 = W(\partial\Omega)$$

and therefore

$$G_{\mu_\Omega} = W(\partial\Omega) G_\mu = W(\partial\Omega) u_\Omega$$

on $R - \bar{R}_0$. Now

$$\int_{\partial\Omega} G_{\mu_\Omega} \, dG^*_{\mu_\Omega} = W(\partial\Omega)^2 \int_{\partial\Omega} u_\Omega \, du^*_\Omega = W(\partial\Omega)^2 D_{\Omega - \bar{R}_0}(u_\Omega).$$

On the other hand, by (g) of Lemma 2F

$$\int_{\partial\Omega} G_{\mu_\Omega} \, dG^*_{\mu_\Omega} = W(\partial\Omega) \int_{\partial\Omega} dG^*_{\mu_\Omega} = W(\partial\Omega) \int_{\partial R_0} d^*_z \int_{\partial\Omega} G(z, t) \, d\mu_\Omega(t)$$

$$= W(\partial\Omega) \int_{\partial\Omega} \left(\int_{\partial R_0} d_z G^*(z, t) \right) d\mu_\Omega(t) = 2\pi W(\partial\Omega).$$

We conclude that $2\pi = W(\partial\Omega) D_{\Omega - \bar{R}_0}(\mu_\Omega)$.

4D. It can be shown that $D(\partial\Omega) = W(\partial\Omega)$, but we only need the following inequality:

Lemma. *The transfinite diameter of $\partial\Omega$ dominates the reciprocal of the capacity,*

$$D(\partial\Omega) \ge W(\partial\Omega).$$

Proof. For each n we can find a system of points $z_{n1}, \cdots, z_{nn}$ in $\partial\Omega$ such that

$$\binom{n}{2} D_n(\partial\Omega) \ge \sum_{i<j}^{1, \cdots, n} G(z_{ni}, z_{nj}) - \frac{1}{n}.$$

Let μ_n be the measure on $\partial\Omega$ such that $\mu_n(z_{ni}) = n^{-1}$ with $i = 1, \cdots, n$ and $S_{\mu_n} \subset \bigcup_{i=1}^n z_{ni}$. Then $\mu_n \in m_{\partial\Omega}$. We can find a subsequence $\{\mu_{n_k}\}$ of $\{\mu_n\}$ and a measure μ such that

$$\lim_{k\to\infty} \int f\, d\mu_{n_k} = \int f\, d\mu$$

for any finitely continuous function f on $\partial\Omega$. For the proof see Halmos [1, pp. 243–249].

It is clear that $\mu \in m_{\partial\Omega}$. For $c > 0$ the function $G_c(z, t) = \min(G(z, t), c)$ is finitely continuous on $\partial\Omega \times \partial\Omega$. By the Stone-Weierstrass approximation theorem (see, e.g., Loomis [1, pp. 9–10]) there exists a function

$$\varphi_n(z, t) = \sum_1^m a_j f_j(z) h_j(t)$$

with a_j real and f_j, h_j finitely continuous on $\partial\Omega$ such that

$$|G_c(z, t) - \varphi_n(z, t)| < \frac{1}{n}.$$

We infer that

$$\text{(10)} \qquad D_{n_k}(\partial\Omega) + \frac{1}{n_k} > \frac{1}{n_k^2} \sum_{i \neq j}^{1, \cdots, n_k} G(z_{n_k i}, z_{n_k j})$$
$$\geq \int\!\!\int G_c\, d\mu_{n_k}\, d\mu_{n_k} - \frac{c}{n_k}$$
$$\geq \int\!\!\int \varphi_n\, d\mu_{n_k}\, d\mu_{n_k} - \frac{1}{n} - \frac{c}{n_k}.$$

Observe that

$$\int\!\!\int \varphi_n\, d\mu_{n_k}\, d\mu_{n_k} = \sum_1^m a_j \int f_j\, d\mu_{n_k} \int h_j\, d\mu_{n_k}$$
$$\to \sum_1^m a_j \int f_j\, d\mu \int h_j\, d\mu = \int\!\!\int \varphi_n\, d\mu\, d\mu$$
$$\geq \int\!\!\int G_c(z, t)\, d\mu(z)\, d\mu(t) - \frac{1}{n}$$

as $k \to \infty$. In (10) by first letting $k \to \infty$ and then $n \to \infty$ we obtain

$$D(\partial\Omega) \geq \int\!\!\int G_c(z, t)\, d\mu(z)\, d\mu(t).$$

Again for $c \to \infty$ we deduce that $D(\partial\Omega) \geq \|\mu\|^2 \geq W(\partial\Omega)$.

4E. We are now able to show (Nakai [1]):

Theorem. *If R is parabolic, then the Čech boundary Γ of R is so small that*

$$E(\Gamma) = D(\Gamma) = \infty.$$

Proof. By Lemmas 3C, 3D, 4D, and 4C

$$E(\Gamma) \geq D(\Gamma) \geq D(\partial\Omega) \geq W(\partial\Omega) \geq \frac{2\pi}{D_{\Omega-\bar{R}_0}(u_\Omega)},$$

where Ω is a regular region with $\bar{R}_0 \subset \Omega$. Since R is parabolic, $u_\Omega \to 0$ as $\Omega \to R$ and consequently $D_{\Omega-\bar{R}_0}(u_\Omega) \to 0$.

5. Construction

5A. We denote by $a_\infty = a_\infty(R)$ the Alexandroff ideal boundary point of R. Then the complements of regular regions form a base for the neighborhoods of a_∞ in $R \cup a_\infty$. First we show (Nakai [1], Kuramochi [3]):

Theorem. *Let R be a parabolic open Riemann surface and let R_0 be a regular region of R with connected complement. Then there exists a positive harmonic function $p(z)$ on $R-\bar{R}_0$ such that*

(a) $\lim_{z\to a_\infty} p(z) = \infty$,

(b) *$p(z)$ has continuously vanishing boundary values on ∂R_0,*

(c) $\int_{\partial R_0} dp^* = 2\pi$, *where ∂R_0 is positively oriented with respect to R_0.*

Proof. By Theorem 4E, $E(\Gamma)=\infty$. Since $E_n(\Gamma) \to E(\Gamma)$, we can find a sequence $\{n_k\}_{k=1}^{\infty}$ such that $E_{n_k}(\Gamma) > 2^{k-1}$. Choose a system of points $x_{k1}, \cdots, x_{kn_k}$ in Γ such that

$$\inf_{x\in\Gamma} \sum_{i=1}^{n_k} G(x, x_{ki}) > n_k 2^{k-1}. \tag{11}$$

We denote by μ_k the measure on Γ with total measure 2^{-k} such that $\mu_k(x_{ki}) = n_k^{-1}2^{-k}$, $i=1, \cdots, n_k$.

Let

$$p_k(z) = \int_\Gamma G(z, y)\, d\mu_k(y) = \sum_{i=1}^{n_k} \frac{G(z, x_{ki})}{n_k 2^k}.$$

Then it is easy to see by Lemma 2F that $p_k(z)$ is positive and harmonic on $R-\bar{R}_0$, vanishes on ∂R_0, and is continuous on $\tilde{R}-R_0$. Hence by (11)

$$p_k(x) > \tfrac{1}{2} \quad \text{on} \quad \Gamma,$$

and we can find a regular region Ω_k in R with $\bar{R}_0 \subset \Omega_k$ such that

$$p_k(z) > \tfrac{1}{2} \quad \text{on} \quad R-\bar{\Omega}_k. \tag{12}$$

Next let $\mu = \sum_1^\infty \mu_k$. Then μ is a unit Borel measure on Γ. Put

$$p(z) = \int_\Gamma G(z, y)\, d\mu(y).$$

By Lemma 2F, $G(z, y)$ is finitely continuous on $(R - R_0) \times \Gamma$; consequently $p(z)$ is continuous on $R - R_0$ and a fortiori on $\tilde{R} - R_0$. Again by Lemma 2F, $p(z)$ vanishes on ∂R_0 and is harmonic on $R - \bar{R}_0$. Thus $p(z)$ satisfies (b).

Clearly

$$p(z) = \sum_1^\infty p_k(z).$$

For arbitrary m let $K_m = \bar{\Omega}_1 \cup \cdots \cup \bar{\Omega}_m$, where Ω_k is as in (12). Then K_m is compact in R with $K_m \supset \bar{R}_0$ and by (12)

$$\frac{m}{2} \leq \sum_1^m p_k(z) \leq p(z)$$

on $R - K_m$. This shows that (a) holds for this $p(z)$. Finally by (g) of Lemma 2F

$$\int_{\partial R_0} dp^* = \int \left(\int_{\partial R_0} d_z G^*(z, q) \right) d\mu(q) = 2\pi.$$

5B. Now let R be an *arbitrary* open Riemann surface. The function $p(z, a)$ on R is called the *Evans-Selberg potential* on R with its pole at $a \in R$ if it satisfies the following three conditions:

(a) *$a \to p(z, a)$ is a harmonic function on $R - a$,*
(b) *$p(z, a) - \log |z - a|$ is harmonic at a,*
(c) $\lim_{z \to a_\infty} p(z, a) = \infty$.

5C. We are ready to state our main result (Nakai [1], Kuramochi [3]):

Theorem. *An open Riemann surface R is parabolic if and only if the Evans-Selberg potential exists on R.*

Proof. First assume that the Evans-Selberg potential $p(z, a)$ exists on R. If we take t sufficiently large, then $D_a = \{z \mid p(z, a) < -t\}$ is a relatively compact disk by (b), (c), and (a). The function $u(z) = p(z, a) + t$ is positive harmonic on $R - \bar{D}_a$, vanishes on ∂D_a, and $\lim_{z \to a_\infty} u(z) = \infty$. Let

$$R_n = \{z \mid z \in R - D_a,\ u(z) < n\} \cup D_a.$$

Then $\{R_n\}_1^\infty$ is an exhaustion of R. Let $u_n(z)$ be harmonic on $\bar{R}_n - D_a$ with $u_n \mid \partial D_a = 0$ and $u_n \mid \partial R_n = 1$. We have $u_n(z) = u(z)/n \to 0$ on $R - D_a$, and R is parabolic (App. I.4).

Conversely suppose R is parabolic and consider the disks R_0: $|z - a| < 2$ and R_1: $|z - a| < 1$. Let $p(z)$ be the function of Theorem 5A. Let L_1 be the operator defined in II.3C for the Riemann surface $R-a$. For the boundary neighborhood take

$$W_1 = (R-\bar{R}_0) \cup (R_1-a)$$

and for the singularity function choose

$$\sigma(z) = \begin{cases} p(z) & \text{in} \quad R-R_0, \\ \log|z-a| & \text{in} \quad \bar{R}_1-a. \end{cases}$$

Then the flux of σ vanishes, i.e.,

$$\int_{\partial W_1} d\sigma^*(z) = \int_{\partial(R-\bar{R}_0)} dp^*(z) + \int_{\partial R_1} d^* \log|z-a| = 0,$$

and by Theorem II.4C there exists a harmonic function $p(z, a)$ on $R-a$ such that

$$L_1(p(z,a)-\sigma(z)) = p(z,a)-\sigma(z).$$

Hence in view of II.(42), $p(z, a)-\sigma(z)$ is bounded on $W_1=(R-\bar{R}_0) \cup (R_1-a)$, and we conclude that (b) and (c) of 5B are satisfied by $p(z, a)$.

The proof of the existence of the Evans-Selberg potential on every parabolic Riemann surface is herewith complete.

§2. MEROMORPHIC FUNCTIONS IN A BOUNDARY NEIGHBORHOOD

The Nevanlinna theory for meromorphic functions in $|z|<\infty$ was first extended by af Hällström [1] and Tsuji [7] to meromorphic functions defined in the complement $\bar{\Omega}-E$ of a compact set E of capacity zero contained in a regular plane region Ω. This was done by using the Evans-Selberg potential for E on the plane as the coordinate in place of $\log|z|$ in the case of $|z|<\infty$. In view of this the existence of the Evans-Selberg potential for parabolic surfaces of arbitrary genus had been sought for some time. Now that we have this potential we can prove the counterpart of the af Hällström-Tsuji theorem for boundary neighborhoods of parabolic surfaces which are thus special cases of R_p-surfaces (cf. I.12). As a consequence we obtain an extension to parabolic surfaces of the af Hällström-Nevanlinna-Kametani theorem on exceptional sets.

6. The af Hällström-Tsuji approach

6A. Let R be a parabolic open Riemann surface and R_1 the complement in R of the closure of a regular region R_0. Here we allow the case $R_1=R$, i.e., $R_0=\varnothing$. Thus R and R_1 have the same ideal boundary β. Fix a point

$z_0 \in R$ and let $p(z, z_0)$ be the Evans-Selberg potential on R with pole z_0 (Theorem 5C). Let $\theta(z, z_0)$ be the *harmonic conjugate* of $p(z, z_0)$, not single-valued in general, and set

$$Z = Z(z, z_0) = e^{p(z,z_0)+i\theta(z,z_0)} = r(z, z_0)e^{i\theta(z,z_0)}.$$

Then $r(z, z_0)$ and $\theta(z, z_0)$ play a role similar to $|z|$ and $\arg z$ in the case of $|z| < \infty$, and Z can be used as a local parameter on R except perhaps at a countable set of isolated points. Hereafter we simply write

$$\begin{cases} r = r(z) = r(z, z_0) \\ \theta = \theta(z) = \theta(z, z_0). \end{cases}$$

We set $r_1 = \inf\{r' \mid \beta_r \subset \bar{R}_1, r \geq r'\}$, where β_r is the "level line"

$$\beta_r = \{z \mid z \in R, r(z, z_0) = r\}, \qquad 0 < r < \infty,$$

positively oriented with respect to $\{z \mid z \in R, r(z, z_0) < r\}$. We also set

$$\Omega_r = \{z \mid z \in R_1, r(z, z_0) < r\}, \qquad r > r_1.$$

Then $\bar{\Omega}_r$, is compact in R, with $\Omega_r \to R_1$ and $\Omega_r \cup \bar{R}_0 \to R$ as $r \to \infty$. Although θ is not single-valued in general,

$$d\theta = d\theta(z, z_0) = dp^*(z, z_0)$$

has a definite meaning regardless of which branch of θ we choose, and

$$\int_{\beta_r} d\theta = 2\pi. \tag{13}$$

6B. Let f be a meromorphic function defined on $\bar{R}_1$. For a point a on the Riemann sphere S: $|\zeta| \leq \infty$ we denote by $n(r, a)$ the number of a-points of f in Ω_r, each counted with its multiplicity. Slightly modifying I.14B we define the counting function by

$$N(r, a) = \int_{r_1}^{r} \frac{n(r, a)}{r} dr$$

and the proximity function by

$$m(r, a) = \frac{1}{2\pi} \int_{\beta_r} \log \frac{1}{[f(z), a]} d\theta,$$

where $[a, b] = |a - b|[(1 + |a|^2)(1 + |b|^2)]^{-1/2}$; then the characteristic function is, by definition,

$$T(r) = \frac{1}{\pi} \int_{r_1}^{r} \frac{A(r)}{r} dr.$$

Here $A(r)$ is the area of the Riemannian image of Ω_r under f on the ζ-sphere S: $|\zeta| \leq \infty$. In the case $r_1 = 0$ we replace $\int_{r_1}^r$ by $\int_{r_1+}^r = \int_{0+}^r$.

Considering f locally as a function of $re^{i\theta}$ and using the argument principle in the same manner as in I.14 we obtain for every meromorphic function in R_1

$$m(r, a) + N(r, a) = T(r) + O(\log r). \tag{14}$$

6C. In the case R_1: $|z| < \infty$, $\lim_{r\to\infty} T(r)/\log r = \infty$ if and only if f has an essential singularity at $z = \infty$. To obtain a similar result in our case we have to define an analogue of an essential singularity at the ideal boundary of R. In view of the Weierstrass theorem we introduce:

Definition. *A meromorphic function f on $R_1 = R - \bar{R}_0$ is said to have the Weierstrass property at the ideal boundary β of R if the global cluster set*

$$C_{R_1}(f) = \bigcap_{r > r_1} \overline{f(R_1 - \bar{\Omega}_r)}$$

at β is total, i.e.,

$$C_{R_1}(f) = S.$$

In the case R_1: $0 < |z - a| < \rho$, f has the Weierstrass property at a if and only if f has an essential singularity at a. A meromorphic function f on a Riemann surface R satisfying the counterpart of the nondegeneracy condition I.(30) has the Weierstrass property at β (Lemma I.6B). The same is true of functions subject to I.(105) on R_p-surfaces.

6D. We shall show (Nakai):

Theorem. *For a meromorphic function f in a boundary neighborhood $\bar{R}_1 = R - R_0$ of a parabolic Riemann surface R*

$$\lim_{r\to\infty} \frac{T(r)}{\log r} = \infty \tag{15}$$

if and only if f has the Weierstrass property at the ideal boundary of R.

Proof. First assume that f has the Weierstrass property. It is easy to see that there exists a point a on S such that $n(r, a) \to \infty$ as $r \to \infty$. Then by (14) and by $m(r, a) \geq 0$ we have for sufficiently large r

$$T(r) + O(\log r) \geq N(r, a) = \int_{r_1}^r \frac{n(r, a)}{r}\, dr$$

$$\geq \int_{\sqrt{r}}^r \frac{n(r, a)}{r}\, dr \geq n(\sqrt{r}, a) \int_{\sqrt{r}}^r \frac{dr}{r} = \frac{n(\sqrt{r}, a)}{2} \log r.$$

Hence $2T(r)/\log r \geq n(\sqrt{r}, a) + O(1)$.

Conversely assume that (15) holds. If $S - C_{R_1}(f) \neq \varnothing$, we can find a point $b \in S - C_{R_1}(f) = \bigcup (S - \overline{f(R_1 - \bar{\Omega}_r)})$. In other words, there exists an r_0 such that $b \in S - \overline{f(R_1 - \bar{\Omega}_{r_0})}$ and a fortiori $b \in S - \overline{f(R_1 - \bar{\Omega}_r)}$ for $r \geq r_0$. From this we see that $n(r, b) = n(r_0, b)$ for $r \geq r_0$. This yields

$$N(r, b) \leq \int_{r_1}^{r} \frac{n(r_0, b)}{r}\, dr = n(r_0, b)(\log r - \log r_1)$$

for $r \geq r_0$. Let d be the distance from b to $\overline{f(R - \bar{\Omega}_{r_0})}$ on the ζ-sphere S. Then $d > 0$ and

$$2\pi m(r, b) = \int_{\beta_r} \log \frac{1}{[f(z), b]}\, d\theta \leq 2\pi \log \frac{1}{d}.$$

Thus $T(r) = O(\log r)$, which contradicts (15).

Remark. The sufficiency of (15) for the Weierstrass property is also implied by I.19A in the metric of I.14B. The necessity of I.(105) in I.19A can be established in the same manner as above.

7. Exceptional sets

7A. We can now prove the following result (Nakai) which complements Lemma I.6B.

Theorem. *Let R be a parabolic Riemann surface. Every meromorphic function on $\bar{R}_1 = R - R_0$ with the Weierstrass property takes every value infinitely often in R_1 except perhaps for a countable union of compact sets of capacity zero.*

Proof. Let $K_n = \{a \mid a \in S,\ n(r, a) \leq n,\ r_1 < r < \infty\}$, $n = 1, 2, \cdots$. It is easy to see that K_n is a closed subset of S. We have to show that K_n is of capacity zero.

If this is not so, then we can find a compact set $K \subset K_n$ such that K is of positive capacity and contained in $W_t: |\zeta| < t < \infty$. Let $g(\zeta, a)$ be the Green's function on W_t. By Lemma 4A there exists a measure μ with $S_\mu \subset K$ and $\mu(K) > 0$ such that

$$\int_K g(\zeta, a)\, d\mu(a) \leq 1$$

on W_t. We define $g(\zeta, a) = 0$ if $\zeta \in S - W_t$. It is clear that there exists a constant c such that

$$\log \frac{1}{[\zeta, a]} \leq g(\zeta, a) + c$$

for $(\zeta, a) \in S \times K$. Thus

$$\int_K \log \frac{1}{[\zeta, a]}\, d\mu(a) = O(1)$$

for $\zeta \in S$. It follows that

$$\int_K m(r, a)\, d\mu(a) = \frac{1}{2\pi} \int_{\partial\Omega_r} \left(\int_K \log \frac{1}{[f(z), a]}\, d\mu(a) \right) d\theta = O(1)$$

and

$$\int_K N(r, a)\, d\mu(a) = \int_{r_1}^{r} \int_K n(r, a)\, d\mu(a) \frac{dr}{r} = O(\log r).$$

By virtue of (14), $T(r) = O(\log r)$, which contradicts Theorem 6D.

7B. The original theorem of af Hällström [1], Nevanlinna [22], and Kametani [2] can be deduced from the preceding theorem:

Theorem. *Let E be a compact set of capacity zero contained in a plane region D_0 and let f be a meromorphic function on the region $D = D_0 - E$ possessing an essential singularity at each point z_0 of E. Then in each neighborhood of any z_0, f takes on every value infinitely often except possibly for a countable union K of compact sets of capacity zero.*

Here K is independent of z_0.

Proof. For $z \in E$ let $U_n(z)$, $n = 1, 2, \cdots$, be an open neighborhood of z with diameter less than n such that $\partial U_n(z) \subset D$ and $\bigcap_1^\infty U_n(z) = z$. On taking $R = S - U_n(z) \cap E$ and $R_1 = U_n(z) - E$ we can apply Theorem 7A to the present f, since it clearly has the Weierstrass property on $U_n(z) \cap E$. Let $K_n(z)$ be the set of values which f takes only a finite number of times in $U_n(z) - E$. Then by Theorem 7A, $K_n(z)$ is a countable union of compact sets of capacity zero. Since $\{U_n(z) \mid z \in E\}$ is an open covering of E, we can find a finite subcovering $\{U_n(z_{ni})\}$, $i = 1, \cdots, k_n$. For each $z \in E$ choose a $U_n(z_{ni_z^n})$ containing z with $1 \le i_z^n \le k_n$. Then $\bigcap_{n=1}^\infty U_n(z_{ni_z^n}) = z$. Thus in each neighborhood of $z_0 \in E$, f assumes every value in $S - \bigcup_{n=1}^\infty \bigcup_{i=1}^{k_n} K_n(z_{ni})$ infinitely often and $K = \bigcup_{n=1}^\infty \bigcup_{i=1}^{k_n} K(z_{ni})$ is a countable union of compact sets of capacity zero.

7C. A value a is said to be a *Picard value of f at $z_0 \in E$* if there exists a neighborhood of z_0 in which f omits a. If E consists of a single point z_0, Picard's Great Theorem states that the number of Picard values of f at z_0 is at most two. In view of this one might suspect that the above theorem could be substantially improved. That it is in fact sharp is a recent result of Matsumoto [1] to be established in the next chapter.

CHAPTER V

PICARD SETS

For the set of Picard values of a meromorphic function at a point z_0, i.e., values omitted by it in a neighborhood of z_0, we shall use in this book the suggestive term Picard set. We know (IV.7B) that the Picard set of a meromorphic function at each point of a set of essential singularities of vanishing capacity is again a set with this property. We shall show in §1 that this result is the best possible: for an arbitrary compact set K of capacity zero there exists a meromorphic function f with exactly K as its Picard set at every essential singularity, the set E of these singularities having vanishing capacity. On the other hand, if E has positive capacity, we can construct an f such that its Picard set at each point of E is of positive capacity even if the linear measure of E vanishes.

Finite Picard sets are considered in §2. A sufficient condition is given for a set E to permit only finite Picard sets at each point of E for any f. This result can be considered a generalization of Picard's Great Theorem. It is illustrated by a Cantor set allowing at most three Picard values.

All results obtained in this chapter are due to Matsumoto [1], [2], and [4]. They are sharpenings of earlier theorems by af Hällström [1], Nevanlinna [22], Kametani [2], Noshiro [4], and Carleson [2].

§1. INFINITE PICARD SETS

We shall first consider sets of vanishing capacity, then those of positive capacity.

1. Sets of capacity zero

1A. A point ζ_0 in the extended ζ-plane is said to be a Picard value of a meromorphic function f at a point z_0 of the z-plane if $f(z) \neq \zeta_0$ for all points z in some neighborhood of z_0. The *Picard set* at z_0 is the union of Picard values at z_0.

We denote by CE the complement of a set E with respect to the extended plane. By "capacity" we shall always mean the logarithmic capacity (App. I.7).

We state the sharpness of Theorem IV.7B (Matsumoto [1]):

Theorem. *For every compact set K of capacity zero in the ζ-plane there exists a compact set E of capacity zero in the z-plane and a meromorphic function f in CE such that f has an essential singularity at each point of E and its Picard set at each singularity coincides with K.*

The proof will be given in 1B to 1D.

1B. Let $\{S_n\}$ be an exhaustion of the region $S=CK$ of the extended ζ-plane, i.e., a sequence of regular regions of S with $\bar{S}_n \subset S_{n+1}$ and $\bigcup_0^\infty S_n = S$. Denote by $N(n)$ the number of components of ∂S_n and by $\omega_n(\zeta)$ the harmonic measure (App. I.2) of ∂S_n with respect to the region $S_n - \bar{S}_0$. Without loss of generality we assume that $N(0)=1$ and that our exhaustion satisfies the condition

$$\lim_{n\to\infty}\left(\prod_{\nu=0}^{n-1}(N(\nu)+1)\right)D_n(\omega_n) = 0, \tag{1}$$

where $D_n(\omega_n)$ is the Dirichlet integral of ω_n over $S_n - \bar{S}_0$. In fact, $\prod_{\nu=0}^{n-1}(N(\nu)+1)$ depends only on $S_0, \cdots, S_{n-1}$ and, since S is parabolic, we can choose S_n so that $D_n(\omega_n)$ is arbitrarily small (App. I.4).

The open set $S_n - \bar{S}_{n-1}$ for $n \geq 1$ consists of $N(n-1)$ components S_{nm} with $m=1, \cdots, N(n-1)$. We take an arbitrary slit L_{nm} in S_{nm} and set

$$S^0 = S^1 = S - \bigcup_{n=1}^{\infty}\bigcup_{m=1}^{N(n-1)} L_{nm},$$

$$S_k^2 = S - \bigcup_{n=3}^{\infty}\bigcup_{m=1}^{N(n-1)} L_{nm} - L_{2k} \qquad \text{with} \quad k = 1, \cdots, N(1),$$

$$\cdots$$

$$S_k^p = S - \bigcup_{n=p+1}^{\infty}\bigcup_{m=1}^{N(n-1)} L_{nm} - L_{pk} \qquad \text{with} \quad k = 1, \cdots, N(p-1),$$

$$\cdots.$$

We shall construct a planar parabolic covering surface $\hat{S}$ of the extended ζ-plane that covers no point in K, but covers all points in CK infinitely often.

First we connect S^1 with S^0 crosswise across the slit $L_{1,1}$ and denote the resulting surface by $\hat{S}^1$. Then $\hat{S}^1$ is a covering surface with $N(0)+1=2$ sheets. Next we take $N(0)+1=2$ replicas of each S_k^2 for $k=1, \cdots, N(1)$ and connect each replica with $\hat{S}^1$ crosswise across the slit L_{2k}. Thus we obtain a $(N(0)+1)(N(1)+1)$-sheeted covering surface $\hat{S}^2$ containing $\hat{S}^1$ as a subsurface.

We continue the construction inductively. Having formed $\hat{S}^n$ with

$\prod_{\nu=0}^{n-1}(N(\nu)+1)$ sheets we take $\prod_{\nu=0}^{n-1}(N(\nu)+1)$ replicas of each S_k^{n+1} for $k=1,\cdots,N(n)$ and connect each of them with $\hat{S}^n$ crosswise across the slit $L_{n+1,k}$ so as to obtain $\hat{S}^{n+1}$ with $\prod_{\nu=0}^{n}(N(\nu)+1)$ sheets. The limiting surface $\hat{S}$ of $\hat{S}^n$ as $n\to\infty$ is a planar covering surface. By construction the set of all points covered by $\hat{S}$ infinitely often is precisely S while no point of K is covered at all.

1C. We proceed to prove that the covering surface $\hat{S}$ is parabolic. Let $\hat{S}_n$ with $n>0$ be the part of $\hat{S}^n\subset\hat{S}$ over the subregion S_n of the base surface S. For $n=0$ we denote by $\hat{S}_0$ the region S_0 on $S^0\subset\hat{S}$. It is easy to see that $\hat{S}_n$ is a relatively compact region of $\hat{S}$ and that $\bar{\hat{S}}_n\subset\hat{S}_{n+1}$ for every $n\geq 0$. Moreover, $\{\hat{S}_n\}_0^\infty$ exhausts $\hat{S}$.

Let $\hat{\omega}_n$ be the harmonic measure of $\partial\hat{S}_n$ with respect to $\hat{S}_n-\bar{\hat{S}}_0$ and form the subharmonic function $\tilde{\omega}_n$ on S_n such that $\tilde{\omega}_n=\omega_n$ on $S_n-\bar{S}_0$ and $\tilde{\omega}_n=0$ on $\bar{S}_0$. Denote by φ the projection of $\hat{S}$ on the extended ζ-plane and by $\hat{D}_n(\tilde{\omega}_n\circ\varphi)$ the Dirichlet integral of $\tilde{\omega}_n\circ\varphi$ over the region $\hat{S}_n-\bar{\hat{S}}_0$. By the Dirichlet principle

$$\hat{D}_n(\hat{\omega}_n)\leq\hat{D}_n(\tilde{\omega}_n\circ\varphi)=\prod_{\nu=1}^{n-1}(N(\nu)+1)D_n(\omega_n),$$

because $\hat{S}_n$ covers each point of S_n exactly $\prod_{\nu=1}^{n-1}(N(\nu)+1)$ times. From this it follows by (1) that $\lim_{n\to\infty}\hat{D}_n(\hat{\omega}_n)=0$, and we conclude that $\hat{S}$ is parabolic (App. I.4).

1D. Since $\hat{S}$ is planar there exists a bijective conformal map $\hat{f}$ of $\hat{S}$ onto a region R of the extended z-plane with $E=CR$ of capacity zero. The function $f=\varphi\circ\hat{f}^{-1}$ satisfies the conditions of the theorem. In fact, let z_0 be a point of E and let r be a positive number such that the circle c: $|z-z_0|=r$ does not meet E. Since the image $\hat{f}^{-1}(c)$ of c on $\hat{S}$ is a compact set, there exists an n such that $\hat{S}_n$ contains $\hat{f}^{-1}(c)$, and the disk $|z-z_0|<r$ contains the image of at least one component of $\hat{S}-\bar{\hat{S}}_n$ under $\hat{f}$. Every component of $\hat{S}-\bar{\hat{S}}_n$ contains as a subregion at least one replica of S_k^m for $k=1,\cdots,N(m-1)$ and $m>n$. It follows that in an arbitrary neighborhood $U(z_0)$ of $z_0\in E$, f takes on each value contained in the complement of K with respect to the extended ζ-plane infinitely often. Thus our theorem is established.

Remark. Modifying the above construction we can also prove the theorem under the assumption that *the given set K is a countable union of compact sets of capacity zero.* For the details of the construction we refer the reader to Matsumoto [1].

As a consequence of this extension we see that Picard sets cannot only be uncountably infinite but may also have considerable extent. Take an uncountable compact set K of capacity zero in the ζ-plane. Let K_ζ be the translation of K with respect to the vector ζ. The union of the sets K_ζ for all rational complex numbers ζ is an example.

2. Sets of positive capacity

2A. We shall now show that even relatively small sets of singularities can permit Picard sets of positive capacity (Matsumoto):

Theorem. *There exists a meromorphic function with a set of essential singularities of vanishing linear measure and with a Picard set of positive capacity at each singularity.*

The proof will be given in 2B to 2D.

2B. The Cantor set on the interval $I_0: [-2^{-1}, 2^{-1}]$ of length 1 is constructed by first centrally removing from I_0 a segment of length $1-\xi_1$ with $0<\xi_1<1$. The remaining set I_1 consists of equal segments $I_{1,1}$ and $I_{1,2}$ of total length ξ_1. Inductively one removes centrally from each segment I_{nk} of I_n with $n=1, 2, \cdots$ and $k=1, \cdots, 2^n$ a segment of length $2^{-n}(1-\xi_{n+1})\prod_1^n \xi_i$ with $0<\xi_{n+1}<1$ so as to obtain a set I_{n+1} of equal segments of total length $\prod_1^{n+1}\xi_i$. It is known (Nevanlinna [22, p. 155]) that the capacity of the Cantor set $\bigcap_1^\infty I_n$ vanishes if and only if

$$\sum_1^\infty \frac{\log \xi_n^{-1}}{2^n} = \infty.$$

The choice $\xi_n = 2l^n$ with a fixed $0<l<1/2$ provides us with segments of equal lengths $l^{n(n+1)/2}$ and with a Cantor set K of positive capacity. Clearly K is of linear measure zero.

2C. Consider the complement S of K with respect to the extended ζ-plane. Choose an n_0 such that

$$(1-l^{n_0+1})l^{n_0(n_0+1)/2} < \tfrac{2}{5}(1-l^{n_0})l^{n_0(n_0-1)/2}$$

and consider for $n \geq n_0$ the annular regions

$$A_{nk} = \{\zeta \mid (1-l^{n+1})l^{n(n+1)/2} < |\zeta-\zeta_{nk}| < \tfrac{2}{5}(1-l^n)l^{n(n-1)/2}\},$$

with ζ_{nk} the midpoint of I_{nk}. The regions have the same modulus (App. I.10) $\log \mu_n = \log [2(1-l^n)/5(1-l^{n+1})l^n]$, and for $n \geq n_0$ the $A_{n+1,k}$ together separate K from the A_{nk}. We suppose that A_{nk} encloses $A_{n+1,2k-1}$ and $A_{n+1,2k}$ and denote by Δ_{nk} the triply connected region bounded by the inner contour of A_{nk} and the outer contours of $A_{n+1,2k-1}$ and $A_{n+1,2k}$.

The intersection of I_{nk} and the closure of Δ_{nk} for $n \geq n_0$ and $k=1, \cdots, 2^n$ is a slit L_{nk} in S. We set

$$S^0 = S - \bigcup_{n \geq n_0} \bigcup_{m=1}^{2^n} L_{nm},$$

$$S_k^1 = S - \bigcup_{n \geq n_0+1} \bigcup_{m=1}^{2^n} L_{nm} - L_{n_0 k} \quad \text{with} \quad k = 1, \cdots, 2^{n_0},$$

$$\cdots$$

$$S_k^p = S - \bigcup_{n \geq n_0+p} \bigcup_{m=1}^{2^n} L_{nm} - L_{n_0+p-1,k} \quad \text{with} \quad k = 1, \cdots, 2^{n_0+p-1},$$

$$\cdots.$$

Using these slit regions, we construct in the same way as in 1B a planar covering surface $\hat{S}$ of the extended ζ-plane that covers no point in K but covers all points in CK infinitely often.

2D. We shall show that $\hat{S}$ is in the class O_{AB} of Riemann surfaces that do not carry nonconstant bounded analytic functions.

Let $\hat{S}^p$ with $p=1, 2, \cdots$ be the subregions of $\hat{S}$ that correspond to the surfaces $\hat{S}^p$ of 1B. The annular regions in $\hat{S}^{p+1}$ lying over $A_{n_0+p+1,k}$ together separate the ideal boundary of $\hat{S}$ from the annular regions in $\hat{S}^p$ lying over the $A_{n_0+p,k}$. The number $\nu(p)$ of these annular regions in $\hat{S}^p$ is

$$2^{n_0+p} \cdot \prod_{\nu=0}^{p-1} (2^{n_0+\nu}+1),$$

since $\hat{S}_p$ has $\prod_{\nu=0}^{p-1} (2^{n_0+\nu}+1)$ sheets. Note that their moduli are $\log \mu_{n_0+p}$.

Choose numbers l_1 and l_2 such that $l < l_1 < l_2 < 1/2$. Then

$$\mu_{n_0+p} = \frac{2(1-l^{n_0+p})}{5(1-l^{n_0+p+1})l^{n_0+p}} > l_1^{-(n_0+p)}$$

and

$$\nu(p) = 2^{n_0+p} \cdot \prod_{\nu=0}^{p-1} (2^{n_0+\nu}+1) < \prod_{\nu=0}^{p_0} (2^{n_0+\nu}+1) \cdot \prod_{\nu=p_0+1}^{p} l_2^{-(n_0+\nu)}$$

$$= A l_2^{-(p-p_0)(2n_0+p_0+p+1)/2}$$

for a suitable p_0 and any $p \geq p_0$, where we have set $A = \prod_{\nu=0}^{p_0}(2^{n_0+\nu}+1)$. It follows that for $N > p_0$

$$\sum_{p=1}^{N} \log \mu_{n_0+p} - \log \nu(N) > \sum_{p=p_0+1}^{N} \log l_1^{-(n_0+p)}$$

$$+ 2^{-1}(N-p_0)(N+2n_0+p_0+1) \log l_2 - \log A$$

$$= 2^{-1}(N-p_0)(N+2n_0+p_0+1) \log \frac{l_2}{l_1} - \log A,$$

so that

$$\lim_{N\to\infty}\left\{\sum_{p=1}^{N}\log\mu_{n_0+p}-\log\nu(N)\right\}=\infty.$$

By the criterion of Pfluger [2]-Mori [2] or of Kuroda (App. I.16–19) we conclude that $\hat{S}$ belongs to the class O_{AB}.

We have taken the slits L_{nk} on the real axis of the ζ-plane. Therefore there exists a bijective conformal mapping $\hat{f}$ of $\hat{S}$ onto the complement R of a compact set E on the real axis with respect to the extended z-plane. Since $\hat{S}$ belongs to the class O_{AB} we conclude by a theorem of Kametani [1] and Ahlfors-Beurling [1] that E is of linear measure zero (cf. Ahlfors-Sario [1, pp. 252–254]).

Denote by φ the projection of $\hat{S}$ into the extended ζ-plane. We see by the same reasoning as in 1D that the function $f=\varphi\circ\hat{f}^{-1}$ satisfies all conditions of our theorem. The proof is herewith complete.

§2. FINITE PICARD SETS

We turn to the following problem: Is there a perfect set E in the z-plane such that every function f meromorphic in the complementary region of E with an essential singularity at each point of E has at most a finite number of Picard values at each singularity?

Recently Carleson [2] and Matsumoto [2], [4] gave an affirmative solution to this problem by establishing sufficient conditions for sets E to have the above property. In particular, Carleson exhibited sets of positive capacity permitting at most three Picard values.

In the sequel we shall derive a sufficient condition, due to Matsumoto [2], [4], that substantially sharpens earlier results.

3. Generalized Picard theorem

3A. Let E be a *perfect set* of the extended z-plane, i.e., a closed set containing no isolated points. Clearly there exists an exhaustion $\{R_n\}$ of the complement R of E with the additional property that each component R_{nm} of $R_n-\bar{R}_{n-1}$ is doubly connected. The regularity of the exhausting regions is modified only in that multiple points of the components of ∂R_n are permitted where R_{nm} branches off into two or more components of $R_{n+1}-\bar{R}_n$. We shall call this exhaustion *normal.*

Set $\alpha_{n-1,m}=\partial R_{nm}\cap\partial R_{n-1}$ and $\beta_{nm}=\partial R_{nm}\cap\partial R_n$. Let u_{nm} be the harmonic function in R_{nm} with boundary values 0 on $\alpha_{n-1,m}$ and a constant μ_{nm} on β_{nm} such that the flux is $\int_{\beta_{nm}}du_{nm}^*=2\pi$. Here β_{nm} is oriented positively with respect to R_{nm}. Similarly let u_n be the harmonic function in $R_n - \bar{R}_{n-1}$ with boundary values zero on ∂R_{n-1} and a constant σ_n on ∂R_n such that $\int_{\partial R_n} du_n^* = 2\pi$.

For a suitable choice of the additive constant of u_n^* the function $u_n + iu_n^*$ maps R_{nm} cut along a level line of u_n^* onto a rectangle $0 < u_n < \sigma_n$, $b_m < u_n^* < a_m + b_m$ with

$$a_m = 2\pi \frac{\sigma_n}{\mu_{nm}}, \qquad \sum_{m=1}^{N(n)} a_m = 2\pi$$

and

$$b_1 = 0, \qquad b_m = \sum_{i=1}^{m-1} a_i$$

with $1 < m \leq N(n)$. Consequently $u_n + iu_n^*$ maps $R_n - \bar{R}_{n-1}$ with suitable slits bijectively onto a slit rectangle $0 < u_n < \sigma_n$, $0 < u_n^* < 2\pi$. We define the function $u + iu^*$ by $u_n + iu_n^* + \sum_{j=0}^{n-1} \sigma_j$ on each $R_n - \bar{R}_{n-1}$, $n \geq 1$, with $\sigma_0 = 0$. It maps $R - \bar{R}_0$ which has at most a countable number of suitable slits onto a strip $0 < u < L$, $0 < u^* < 2\pi$ with

$$L = \sum_{j=1}^{\infty} \sigma_j \leq \infty.$$

This slit strip is the *graph* of R associated with the exhaustion $\{R_n\}$ in the sense of Noshiro [5]. The number L is called the length of this graph. By theorems of Sario [2] and Noshiro [5], R is the complement of a compact set of capacity zero if and only if there exists a graph of R whose length L is infinite (App. I.14–15).

3B. Let β_r be the level line $\{z \mid u(z) = r, 0 < r < L\}$ on R. Except for a countable set of values r which we shall exclude, β_r consists of a finite number of Jordan curves β_{rm}, $m = 1, 2, \cdots, n(r)$. We shall call each component of the open set $R_n - \bar{R}_k$, $n > k \geq 0$, an R-chain. For every β_{rm} consider the longest doubly connected R-chain $R(\beta_{rm})$ such that β_{rm} is contained in $R(\beta_{rm})$ or is one of the two components of $\partial R(\beta_{rm})$. Denote by $\log \mu(\beta_{rm})$ the modulus of this R-chain and set

$$\mu(r) = \min_{1 \leq m \leq n(r)} \mu(\beta_{rm}).$$

Generally R_{nm} may branch off at β_{nm} into a finite number of regions $R_{n+1,k}$. If every R_{nm} branches off into at most ρ regions $R_{n+1,k}$, we say that the exhaustion $\{R_n\}$ branches off at most ρ times everywhere.

3C. We are ready to state our main criterion (Matsumoto [2], [4]):

Theorem. *Let E be a totally disconnected compact set in the z-plane and let R be its complementary region. If there exists a normal exhaustion $\{R_n\}$ of R which branches off at most ρ times everywhere and if*

$$\lim_{r \to L} \mu(r) = \infty, \tag{2}$$

then every function meromorphic in R with an essential singularity at each point of E has at most $\rho+1$ Picard values at each singularity.

3D. We shall actually prove more. Let f be meromorphic in R with at least one essential singularity in E, not necessarily at each point of E. We shall say that f has a Picard value ζ in R at an essential singularity $z_0 \in E$ if there is a neighborhood $U(z_0)$ of z_0 such that $f(z) \neq \zeta$ in $R \cap U(z_0)$. Such a value ζ could be taken by f at points of E near z_0 where f is meromorphic.

We claim (Matsumoto [4]):

Theorem. *Under the same conditions on R as in Theorem* 3C, *every function meromorphic in R with at least one essential singularity in E has at most $\rho+1$ Picard values in R at each singularity.*

If a meromorphic function in R omits $\rho+2$ values in R it must consequently be rational.

In the case $\rho=1$ the set E reduces to a point and our assertion is true by Picard's theorem. We shall give the proof for $\rho=2$. The general case is analogous if we take $\rho+2$ points ζ_i instead of 4 in Lemma 4A below.

4. Auxiliary results

4A. Before proving the theorem we shall establish two lemmas.

Consider the Riemann sphere Σ with radius $1/2$ tangent to the ζ-plane at the origin. For ζ and ζ' in the extended ζ-plane we denote by $[\zeta, \zeta']$ their chordal distance. Further let $C(\zeta; \delta)$ with $\delta > 0$ be the spherical open disk with center ζ and chordal radius δ.

Let f be a meromorphic function in an annulus $1 < |z| < e^{\sigma}$ omitting four values ζ_i, $i = 1, \cdots, 4$, and let $\delta > 0$ be so small that the spherical disks $C(\zeta_i; \delta)$ are disjoint by pairs.

We recall the Bohr-Landau theorem (e.g., Tsuji [19, p. 268]):

If g is analytic in $|z| < 1$ and $g(z) \neq 0, 1$ there, then

$$\max_{|z|=r} |g(z)| \leq \exp\left(\frac{D \log(|g(0)|+2)}{1-r}\right)$$

for all $r < 1$, with D a positive constant independent of g. Using this inequality (a precise form of Schottky's theorem) we shall establish the following auxiliary result:

Lemma. *There is a positive constant $\delta' > 0$ such that, if f takes a value outside of $C(\zeta_i;\ \delta)$ for some i at a point on $|z| = e^{\sigma/2}$, then the image of $|z| = e^{\sigma/2}$ under f lies completely outside the concentric spherical disk $C(\zeta_i;\ \delta')$. Here δ' depends only on σ, δ, and the points ζ_i, not on f.*

Proof. By the Bohr-Landau theorem we see that if g is an analytic function in $1 < |z| < e^{\sigma}$ such that

$$g(z) \neq 0, 1 \quad \text{and} \quad \min_{|z| = e^{\sigma/2}} |g(z)| < M \quad \text{for some} \quad M > 0,$$

then there is a constant $M' > 0$ depending only on M and σ and satisfying the inequality

$$\max_{|z| = e^{\sigma/2}} |g(z)| \leq M'.$$

Denote by $t = T^i_{jk}(\zeta)$ with distinct i, j, k the linear transformation which maps ζ_i, ζ_j, and ζ_k to the point at infinity, the origin, and the point $t = 1$, respectively. Since the number of such T^i_{jk} is finite, there exists a positive M so large that for each T^i_{jk} the image of the exterior of $C(\zeta_i;\ \delta)$ is contained in $|t| < M$. For this M, $t = T^i_{jk}(f(z))$ has the same properties as g stated above, and hence

$$|T^i_{jk}(f(z))| \leq M' \quad \text{on} \quad |z| = e^{\sigma/2}$$

with $M' > 0$ depending only on M and σ. The image of the exterior V of $|t| \leq M'$ under $(T^i_{jk})^{-1}$ is an open disk containing ζ_i. Denote by d^i_{jk} the chordal distance between ζ_i and the boundary of $(T^i_{jk})^{-1}(V)$. The minimum under all distinct i, j, k

$$\delta' = \min d^i_{jk}$$

is positive and obviously satisfies all conditions of the lemma.

4B. The following estimate is a revised form of Carleson's [2].

Lemma. *Let f be a meromorphic function in an annulus $1 \leq |z| \leq e^{\mu}$. If f takes no value in a spherical disk $C(\zeta_0;\ \delta)$, then there exists a positive constant A depending only on δ such that the diameter of the image of $|z| = e^{\mu/2}$ under f in terms of the chordal distance is dominated by $Ae^{-\mu/2}$ for sufficiently large μ.*

In particular, if δ is sufficiently close to 1, i.e., the spherical disk $C(-1/\bar{\zeta}_0;\ d)$ complementary to $C(\zeta_0;\ \delta)$ has a sufficiently small radius d, then

$$A < Bd,$$

where B is a positive constant.

Proof. We may assume without loss of generality that the center ζ_0 of $C(\zeta_0;\ \delta)$ is the point at infinity, for otherwise we can map ζ_0 to this point by the linear transformation $(1 + \bar{\zeta}_0\zeta)/(\zeta - \zeta_0)$, under which the chordal distance remains invariant. Let $|\zeta| > M$ be the region in the ζ-plane corresponding to $C(\zeta_0; \delta)$. Then

$$|f(z)| \leq M \quad \text{on} \quad 1 \leq |z| \leq e^{\mu}.$$

By Cauchy's integral formula we have

$$f'(z) = \frac{1}{2\pi i}\left\{\int_{|t|=e^{\mu}} \frac{f(t)}{(t-z)^2}\,dt - \int_{|t|=1} \frac{f(t)}{(t-z)^2}\,dt\right\}$$

for every z on $|z| = e^{\mu/2}$ and hence, if $\mu \geq 2$,

$$|f'(z)| \leq \frac{M}{2\pi}\left\{\frac{2\pi e^{\mu}}{(e^{\mu}-e^{\mu/2})^2} + \frac{2\pi}{(e^{\mu/2}-1)^2}\right\} \leq \frac{2e^2}{(e-1)^2}\,Me^{-\mu}.$$

It follows that the length of $|z| = e^{\mu/2}$ under f has the bound

$$\int_{|z|=e^{\mu/2}} |f'(z)|\,|dz| \leq \frac{2e^2}{(e-1)^2}\,Me^{-\mu}\cdot 2\pi e^{\mu/2} = \frac{4\pi e^2}{(e-1)^2}\,Me^{-\mu/2}.$$

We conclude that the diameter of the image of $|z| = e^{\mu/2}$ with respect to the metric $|d\zeta|$, and consequently with respect to the chordal distance, is dominated by $2\pi e^2(e-1)^{-2}Me^{-\mu/2}$. We can choose $A = 2\pi e^2(e-1)^{-2}M$ as M depends only on δ.

If $d < 1/2$, then $M < 2d$ and

$$B = \frac{4\pi e^2}{(e-1)^2}$$

satisfies our condition. Our lemma is herewith established.

5. Proof of the generalized Picard theorem

5A. Theorem 3D will be proved by contradiction. Suppose there exists a meromorphic function f in R with at least one essential singularity in E and with more than three Picard values at an essential singularity $z_0 \in E$. Then there is a neighborhood $U(z_0)$ of z_0 such that f omits four values ζ_i, $i = 1, \cdots, 4$, in $U(z_0) \cap R$. We take a positive δ so small that the spherical disks $C(\zeta_i; \delta)$ are disjoint by pairs. For this δ and a $\sigma > 0$, Lemma 4A determines $\delta' > 0$.

We take this δ' as δ of Lemma 4B and choose μ_0 so large that

$$Ae^{-\mu_0/2} < \min\left(\frac{1}{24}, \frac{\delta'}{3}\right) \quad \text{and} \quad Be^{-\mu_0/2} < \frac{1}{12},$$

where A and B are the constants of Lemma 4B. By our assumption $\lim_{r \to L} \mu(r) = \infty$ there is an $r_0 > 0$ such that

$$\mu(r) > \mu_0 + 2\sigma \quad \text{for all} \quad r \quad \text{with} \quad r_0 < r < L.$$

5B. The level line $\beta_r=\{z\,|\,u(z)=r\}$ consists of a finite number of Jordan curves β_{rk} with $k=1,\cdots,n(r)$, and one of them, say $\beta_{r,1}$, encloses z_0. For r sufficiently near L the longest doubly connected R-chain $R(\beta_{r,1})=D_{1,1}$ defined in 3B is contained in $U(z_0)$. The modulus of $D_{1,1}$ is greater than μ_0+2 but is not infinite, for otherwise z_0 would have to be isolated and f could not have four Picard values at z_0. Therefore $D_{1,1}$ must branch off. Now suppose $D_{1,1}$ is a component of the open set $R_n-\bar{R}_{n'}$ with $n>n'$, and branches off into two regions $R_{n+1,m}$ and $R_{n+1,m'}$. Consider the longest doubly connected R-chains $D_{2,1}$ and $D_{2,2}$ containing $R_{n+1,m}$ and $R_{n+1,m'}$, respectively. They both have moduli greater than μ_0+2 and one of them, say $D_{2,1}$, separates z_0 from $D_{1,1}$. Its modulus is finite for the same reason as above. Hence $D_{2,1}$ is a component of the open set $R_{\tilde{n}}-\bar{R}_n$ for some $\tilde{n}$ and branches off into two regions $R_{\tilde{n}+1,m}$ and $R_{\tilde{n}+1,m'}$. We denote by $D_{3,1}$ and $D_{3,2}$ the longest doubly connected R-chains containing them. If the modulus of $D_{2,2}$ is infinite, one of the boundary components of $D_{2,2}$ is a point $z_1\in E$ and f is meromorphic at z_1. If the modulus is finite we obtain two R-chains $D_{3,3}$ and $D_{3,4}$ in the same manner as above. Thus we have at most 2^2 R-chains $D_{3,q}$ such that their harmonic moduli are greater than μ_0+2, and one of them encloses z_0. Moreover, each of them branches off into two regions if the modulus is finite, or has a point $z_1\in E$ as one of its boundary components at which f is meromorphic if the modulus is infinite.

Continuing inductively we obtain a set of R-chains D_{pq} with $p=1,2,\cdots$ and $q=1,\cdots,Q(p)\le 2^{p-1}$, which has the following properties:

(a) $\bigcup_{p=1}^{\infty}\bigcup_{q=1}^{Q(p)}\bar{D}_{pq}\supset\Delta$, where Δ is the intersection of R with the set bounded by the Jordan curve $\beta_{r,1}$,

(b) the modulus of each D_{pq} is greater than μ_0+2,

(c) each D_{pq} branches off into two $D_{p+1,q}$ if its modulus is finite,

or

(c′) each D_{pq} has a point $z_1\in E$ as one of its boundary components and f is meromorphic at z_1 if the modulus of D_{pq} is infinite. In this case we shall denote the point z_1 by z_{pq} and the value $f(z_{pq})$ by ζ_{pq}.

5C. Each D_{pq} is conformally equivalent to the annulus $1<|t|<e^{\mu}$. If $\mu<\infty$ we denote by D_{pq}^1, D_{pq}^2, and D_{pq}^3 the subregions of D_{pq} corresponding to the annuli $1<|t|<e^{\sigma}$, $e^{\sigma}<|t|<e^{\mu-\sigma}$ and $e^{\mu-\sigma}<|t|<e^{\mu}$, respectively, and by β_{pq}^1, β_{pq}^2, and β_{pq}^3 the closed curves corresponding to $|t|=e^{\sigma/2}$, $|t|=e^{\mu/2}$, and $|t|=e^{\mu-\sigma/2}$, respectively.

We shall see that for each β_{pq}^2 the diameter of its image under f with respect to the chordal distance is dominated by $K=\min(1/24,\delta'/3)$.

In fact, for $z'\in\beta_{pq}^1$ and $z''\in\beta_{pq}^3$ the images $f(z')$ and $f(z'')$ lie outside of at least one $C(\zeta_i;\delta)$, say $C(\zeta_1;\delta)$, and hence on applying Lemma 4A to D_{pq}^1 and D_{pq}^3 we see that the images of β_{pq}^1 and β_{pq}^3, and, as a consequence of the maximum principle, the image of the annular region bounded by them, lies outside of $C(\zeta_1; \delta')$. Our assertion follows from Lemma 4B, because the modulus of D_{pq}^2 is greater than μ_0.

5D. Every $D_{p+1,q'}$ with $p \geq 1$ has in common with another $D_{p+1,q''}$ a $D_{p,q}$ branching off into them, and we shall denote by Δ_{pq} the triply connected region bounded by β_{pq}^2, $\beta_{p+1,q'}^2$, and $\beta_{p+1,q''}^2$, where $\beta_{p+1,q'}^2 = z_{p+1,q'}$ or $\beta_{p+1,q''}^2 = z_{p+1,q''}$ if $D_{p+1,q'}$ or $D_{p+1,q''}$ has infinite modulus. For $\zeta \in f(\beta_{pq}^2)$, $\zeta' \in f(\beta_{p+1,q'}^2)$, and $\zeta'' \in f(\beta_{p+1,q''}^2)$ we consider spherical disks $C(\zeta; K)$, $C(\zeta'; K)$, and $C(\zeta''; K)$, which of course contain $f(\beta_{pq}^2)$, $f(\beta_{p+1,q'}^2)$, and $f(\beta_{p+1,q''}^2)$, respectively. Since $K < \delta'/3$, there is at least one ζ_i, say ζ_1, not contained in the disks, and we conclude that each disk meets the union of the other two disks. In fact, if this were not the case, there would exist a $z^* \in \Delta_{pq}$ such that $f(z^*)$ can be joined to ζ_1 by a curve Λ in the exterior of the union of these three disks. We would be led to the contradiction that the element of the inverse function f^{-1} corresponding to z^* can be continued analytically along Λ up to a point arbitrarily near to ζ_1 so that f takes the value ζ_1 in Δ_{pq}. We conclude:

(α) For every Δ_{pq} there is a spherical disk with the chordal radius $3K$ containing its image $f(\Delta_{pq})$.

Next consider β_{pq}^2 for $p \geq 2$. The region Δ_{pq} and some $\Delta_{p-1,q'}$ have β_{pq}^2 as the common boundary, and $D_{pq} \subset \Delta_{pq} \cup \beta_{pq}^2 \cup \Delta_{p-1,q'}$. In view of ($\alpha$) the images of $\Delta_{pq} \cup \beta_{pq}^2 \cup \Delta_{p-1,q'}$ and consequently of $D_{pq}^2 \subset D_{pq}$ are contained in a spherical disk with chordal radius $6K < 1/2$. On applying Lemma 4B to D_{pq}^2 for $d = 6K$ we see that the diameter of $f(\beta_{pq}^2)$ is $< 6KBe^{-\mu_0/2} < K/2$. For $p \geq 2$ each boundary component of Δ_{pq} thus has an image with diameter less than $K/2$. By the same reasoning as above we infer:

(β) For $p \geq 2$ the image of every Δ_{pq} is contained in a spherical disk with chordal radius $3K/2$.

By induction we deduce for every n:

(γ) For $p \geq n$ the image of every Δ_{pq} is contained in a spherical disk with chordal radius $3K/2^{n-1}$.

5E. Let Δ' be the intersection of R and the region bounded by the Jordan curve $\beta_{1,1}^2$ and let z^* be a point of $\beta_{1,1}^2$. Then it follows from property (a) of $\{D_{pq}\}$ that

$$\Delta' \subset \bigcup_{p=1}^{\infty} \bigcup_{q=1}^{Q(p)} \overline{\Delta}_{pq},$$

and consequently for any $z' \in \Delta'$ there is a $\Delta_{p'q'}$ whose closure contains z'. From (γ) we have for a chain of Δ_{pq} joining $\Delta_{1,1}$ and $\Delta_{p'q'}$,

$$[f(z'), f(z^*)] \leq \sum_{p=1}^{p'} \operatorname{diam} f(\Delta_{pq}) < 2 \sum_{p=1}^{\infty} \frac{3K}{2^{p-1}} = 12K < \frac{1}{2},$$

where the diameter is in terms of the chordal distance. By means of a linear transformation we conclude that f is bounded in Δ'. On the other hand, on applying the criterion of Pfluger [2]-Mori [2] or of Kuroda (App. I.16–19) to the annular regions $\{D_{pq}\}$ we see easily that the part E' of E contained in the region bounded by $\beta_{1,1}^2$ has a complement of class O_{AB}. Hence each point of E' must be a removable singularity of a bounded function f. This contradicts our assumption that $z_0 \in E'$ is an essential singularity of f, and we conclude that f cannot omit four values in R at z_0.

This completes the proof of Theorem 3D.

5F. For Cantor sets we have the following immediate consequence:

Corollary. *Let E be a Cantor set on the closed interval* [0, 1] *with successive ratios ξ_n satisfying the condition*

$$\lim_{n\to\infty} \xi_n = 0. \tag{3}$$

Then every function meromorphic in $R = CE$ with at least one essential singularity in E has at most three Picard values in R at each singularity.

This result, due to Matsumoto [4], is somewhat sharper than the original theorem of Carleson [2], where the condition

$$\lim_{n\to\infty} \frac{\log \xi_n^{-1}}{\log n} = \infty$$

was used. It is significant that in both cases the capacity of E may be positive, i.e., the extra condition $\sum_1^\infty 2^{-n} \log \xi_n^{-1} < \infty$ can be imposed on E (cf. 2B).

6. Classes of sets with the Picard property

6A. We shall denote by $\mathscr{E}_n$, $n = 1, 2, \cdots$, the class of totally disconnected compact sets E in the z-plane such that every meromorphic function in CE with E as the set of essential singularities has at most $n+1$ Picard values at each singularity. It is clear that

$$\mathscr{E}_1 \subset \mathscr{E}_2 \cdots \subset \mathscr{E}_n \subset \cdots.$$

We claim that these inclusions are strict (Matsumoto [2]):

Theorem. *There exists a set E in the class $\mathscr{E}_n$ such that CE carries a meromorphic function with exactly $n+1$ exceptional values at each singularity.*

By Picard's theorem the set E consisting of a single point meets the requirement of the above theorem for $n=1$. We shall give an example for $n=2$. The construction is similar in the case $n>2$.

6B. We delete the origin and the point $\zeta=1$ from the ζ-plane and denote the resulting region by S. By induction we shall construct covering surfaces $\hat{S}^n$ of the ζ-plane and define an exhaustion $\{\hat{S}_k\}$ of their limiting surface $\hat{S}$.

Take a set $\{A, B; C, D, E\}_\infty$ of analytic Jordan curves satisfying the following conditions with respect to the points ∞, 0, and 1:

(a) A and B each separate ∞ from $0 \cup 1$,

(b) A is tangent to B at one point,

(c) C separates A from ∞,

(d) D encircles 0 and E encircles 1,

(e) D and E are tangent to each other, forming with B the boundary of a doubly connected region $(B, D \cup E)$.

Here we are using the general notation (C_1, C_2) for a doubly connected region with contours C_1, C_2.

We introduce sets $\{A, B; C, D, E\}_0$ and $\{A, B; C, D, E\}_1$ defined as above after cyclically permuting ∞, 0, 1 in (a), (c), and (d).

We also consider a set $\{F, G; H, I\}_\infty$ of analytic Jordan curves with the following properties:

(α) F separates ∞ from $0 \cup 1$,

(β) G is homotopic to zero with respect to S and is tangent to F at one point,

(γ) H separates F from ∞, while I separates H from ∞.

Sets $\{F, G; H, I\}_0$ and $\{F, G; H, I\}_1$ are again defined by cyclically permuting ∞, 0, 1 in (α) and (γ).

6C. First take a replica $\hat{S}^1$ of S. There exists a set

$$\{\alpha_{1,1}, \alpha_{1,2}; \alpha_{2,1}, \alpha_{2,2}, \alpha_{2,3}\}_\infty$$

such that the moduli of the doubly connected regions

$$(\alpha_{1,1}, \alpha_{2,1}) \text{ and } (\alpha_{1,2}, \alpha_{2,2} \cup \alpha_{2,3})$$

are not less than 2. In fact, first choose the curves $\alpha_{2,2}$ and $\alpha_{2,3}$, then $\alpha_{1,2}$ with

$$\operatorname{mod}(\alpha_{1,2}, \alpha_{2,2} \cup \alpha_{2,3}) \geq 2.$$

Next select $\alpha_{1,1}$ and $\alpha_{2,1}$ such that

$$\operatorname{mod}(\alpha_{1,1}, \alpha_{2,1}) \geq 2.$$

The region bounded by $\alpha_{1,1} \cup \alpha_{1,2}$ is taken as $\hat{S}_1$ and the region bounded by $\alpha_{2,1} \cup \alpha_{2,2} \cup \alpha_{2,3}$ as $\hat{S}_2$. Choose $\hat{S}_0$ with $\bar{\hat{S}}_0 \subset \hat{S}_1$ and such that $\hat{S}_1 - \bar{\hat{S}}_0$ is a doubly connected region with modulus ≥ 1.

Next take three replicas S_i, $i=1, 2, 3$, of S. Draw $\{\alpha_{3,1}, \alpha_{3,2}; \alpha_{4,1}, \alpha_{5,1}\}_\infty$ in $\hat{S}^1$ and

$$\{\alpha_{4,2}, \alpha_{4,3}; \alpha_{5,2}, \alpha_{5,3}, \alpha_{5,4}\}_\infty$$

in S_1 as follows. First take $\alpha_{4,3}$, $\alpha_{5,3}$, and $\alpha_{5,4}$ in S_1 with

$$\operatorname{mod}(\alpha_{4,3}, \alpha_{5,3} \cup \alpha_{5,4}) \geq 5.$$

Then determine $\{\alpha_{3,1}, \alpha_{3,2}; \alpha_{4,1}, \alpha_{5,1}\}_\infty$ in $\hat{S}^1$ such that $\alpha_{3,1} \cup \alpha_{3,2}$ is contained in the end part of $\hat{S}^1$ bounded by $\alpha_{2,1}$ and does not intersect the same curve as $\alpha_{4,3}$ drawn in $\hat{S}^1$. Moreover, we require that

$$\operatorname{mod}(\alpha_{2,1}, \alpha_{3,1} \cup \alpha_{3,2}) \geq 3, \quad \operatorname{mod}(\alpha_{3,1}, \alpha_{4,1}) \geq 4,$$

and

$$\operatorname{mod}(\alpha_{4,1}, \alpha_{5,1}) \geq 5.$$

Last trace $\alpha_{4,2}$ and $\alpha_{5,2}$ such that the region bounded by $\alpha_{4,2} \cup \alpha_{4,3}$ contains the same curve as $\alpha_{3,2}$ drawn in S_1 and that

$$\operatorname{mod}(\alpha_{4,2}, \alpha_{5,2}) \geq 5.$$

We connect S_1 with $\hat{S}^1$ crosswise across a slit in the region bounded by $\alpha_{3,2}$. If we choose this slit sufficiently small, we have

$$\operatorname{mod}(\alpha_{3,2}, \alpha_{4,2} \cup \alpha_{4,3}) \geq 4.$$

In a similar manner we draw $\{\alpha_{3,3}, \alpha_{3,4}; \alpha_{4,4}, \alpha_{5,5}\}_0$ and

$$\{\alpha_{3,5}, \alpha_{3,6}; \alpha_{4,7}, \alpha_{5,9}\}_1$$

in $\hat{S}^1$, $\{\alpha_{4,5}, \alpha_{4,6}; \alpha_{5,6}, \alpha_{5,7}, \alpha_{5,8}\}_0$ in S_2, and $\{\alpha_{4,8}, \alpha_{4,9}; \alpha_{5,10}, \alpha_{5,11}, \alpha_{5,12}\}_1$ in S_3. We connect S_2 and S_3 with $\hat{S}^1$ across suitable slits in regions bounded by $\alpha_{3,4}$ and $\alpha_{3,6}$. The resulting surface is denoted by $\hat{S}^2$.

We take as $\hat{S}_3$ the region of $\hat{S}^2$ bounded by $\bigcup_{i=1}^{6} \alpha_{3,i}$, as $\hat{S}_4$ that bounded by $\bigcup_{i=1}^{9} \alpha_{4,i}$, and as $\hat{S}_5$ that bounded by $\bigcup_{i=1}^{12} \alpha_{5,i}$.

6D. Suppose that $\hat{S}^n$ and $\hat{S}_k$ for $0 \leq k \leq 3n-1$ are constructed with $\hat{S}^n$ consisting of 4^{n-1} sheets, and $\partial \hat{S}_{3n-1}$ of $3 \cdot 4^{n-1}$ analytic Jordan curves $\alpha_{3n-1,i}$ each separating one of the three points from the other two. Furthermore, suppose each component of $\hat{S}_{k+1} - \bar{\hat{S}}_k$, which is a doubly connected region, has a modulus not less than $k+1$ for $0 \leq k \leq 3n-2$. Then we take $3 \cdot 4^{n-1}$ replicas S_i with $1 \leq i \leq 3 \cdot 4^{n-1}$ of S and connect each S_i with $\hat{S}^n$ crosswise across a suitable slit in the end part of $\hat{S}^n$ bounded by $\alpha_{3n-1,i}$ as follows.

It suffices to consider only the case where $\alpha_{3n-1,i}$ encircles the point at infinity. In the other cases we simply replace ∞ by 0 or 1. In the same manner as above we choose $\{\alpha_{3n,2i-1}, \alpha_{3n,2i}; \alpha_{3n+1,3i-2}, \alpha_{3n+2,4i-3}\}_\infty$ in the end part of $\hat{S}^n$ bounded by $\alpha_{3n-1,i}$ and

$$\{\alpha_{3n+1,3i-1}, \alpha_{3n+1,3i}; \alpha_{3n+2,4i-2}; \alpha_{3n+2,4i-1}, \alpha_{3n+2,4i}\}_\infty$$

in S_i such that the moduli of the doubly connected regions

$$(\alpha_{3n+1,3i-2}, \alpha_{3n+2,4i-3}), \quad (\alpha_{3n+1,3i-1}, \alpha_{3n+2,4i-2}),$$

and

$$(\alpha_{3n+1,3i}, \alpha_{3n+2,4i-1} \cup \alpha_{3n+2,4i})$$

are not less than $3n+2$, and that

$$\text{mod}\,(\alpha_{3n,2i-1}, \alpha_{3n+1,3i-2}) \geq 3n+1,$$
$$\text{mod}\,(\alpha_{3n-1,i}, \alpha_{3n,2i-1} \cup \alpha_{3n,2i}) \geq 3n.$$

Then we connect S_i with $\hat{S}^n$ crosswise across a slit in the region bounded by $\alpha_{3n,2i}$, choosing it so small that

$$\text{mod}\,(\alpha_{3n,2i}, \alpha_{3n+1,3i-1} \cup \alpha_{3n+1,3i}) \geq 3n+1.$$

On the surface $\hat{S}^{n+1}$ thus obtained we take $\hat{S}_{3n}$, $\hat{S}_{3n+1}$, and $\hat{S}_{3n+2}$ as the regions bounded by $\bigcup_i \alpha_{3n,i}$, $\bigcup_i \alpha_{3n+1,i}$, and $\bigcup_i \alpha_{3n+2,i}$, respectively. It is easily seen that $\hat{S}^{n+1}$ and $\hat{S}_k$ with $0 \leq k \leq 3n+2$ possess all properties listed above for $\hat{S}^n$ and $\hat{S}_k$, $0 \leq k \leq 3n-1$, with n replaced by $n+1$. The limiting surface $\hat{S}$ is a planar covering surface of the extended ζ-plane.

6E. We map $\hat{S}$ bijectively and conformally onto the complement R of a totally disconnected compact set E in the extended z-plane. Let $\hat{f}$ be the mapping function and φ the projection of $\hat{S}$ into the extended ζ-plane. By the same argument as in 1D we see that $f = \varphi \circ \hat{f}^{-1}$ is meromorphic in R, has an essential singularity at each point of E, and possesses three Picard values, 0, 1, and ∞, at each singularity.

Moreover, E satisfies the conditions of Theorem 3C. In fact, the exhaustion of R by the regions $R_k = \hat{f}^{-1}(\hat{S}_k)$ with $k = 0, 1, \cdots$ is clearly normal in the sense of 3A and branches off at most twice everywhere. Furthermore, the moduli of the components of $R_k - \bar{R}_{k-1}$, $k \geq 1$, dominate k, and we have

$$\lim_{r \to L} \mu(r) = \infty,$$

where L is the length of the graph associated with $\{R_k\}$.

6F. The set E thus constructed for $n > 1$ is a perfect set. In this context the question arises: Does $\mathscr{E}_1$ contain any perfect sets? This can be answered affirmatively by a very recent result of Matsumoto [5] (cf. Corollary 5F):

If $\xi_{n+1} = o(\xi_n^2)$, then every function meromorphic in CE with E as the set of essential singularities has at most two Picard values at each singularity.

CHAPTER VI

RIEMANNIAN IMAGES

A milestone in the evolution of value distribution theory was Ahlfors' [11] discovery that the main theorems are not based on the analyticity of the mapping functions but can, in essence, be derived from purely metric-topological properties of covering surfaces. We begin the present chapter by following rather closely Ahlfors' original proof arrangement, unsurpassed in elegance by other presentations. The theory culminates in his fundamental inequality, often referred to as the nonintegrated form of the second main theorem on meromorphic functions. We shall also give this inequality and Picard's theorem as localized by Noshiro [2] to a transcendental singularity of the inverse function.

We then present a generalization by Noshiro [4], [6], and Sario [7] of the nonintegrated form of the second main theorem to mappings of arbitrary Riemann surfaces into closed Riemann surfaces. For surfaces R_p carrying capacity functions with compact level lines (I.12 and IV) we show, following Noshiro-Sario [1], that an integrated form can be derived from this nonintegrated form.

The chapter closes with Noshiro's recent results on algebroids and Sario's [7] and Rodin's [2] examples to prove the sharpness of the "nonintegrated" bound for the number of Picard points.

The presentation is divided into two parts: §§1 to 3 deal with the general theory of orientable covering surfaces, while §§4 and 5 are devoted to Riemannian images under analytic mappings of the plane and of arbitrary Riemann surfaces.

For earlier literature on the subject we refer the reader to Dufresnoy [11], Hayman [4], Kunugui [3], Nevanlinna [22], Tamura [1], Tôki [2], Tsuji [19], Tumura [4], Valiron [46], and others listed in the Bibliography.

§1. MEAN SHEET NUMBERS

After some preliminaries on lengths and areas we introduce the mean sheet numbers of covering surfaces and derive Ahlfors' covering theorems both for regions and curves.

1. Base surface

1A. By a *finite surface* we shall understand a triangulated closed or compact bordered surface. We also fix a sequence of successive subdivisions of the given triangulation. In 1A to 1C we shall consider the former, in 1D, the latter case.

Given a closed triangulated surface S_0 we endow it with a metric satisfying the following conditions:

(a) *There exists a set $\{\gamma\}$ of simple curves γ on S_0 such that to each curve γ there corresponds a finite positive number $|\gamma|$ as its length. Furthermore, $\{\gamma\}$ contains every curve which appears in the triangulation of S_0 and in the successive subdivisions of it.*

(b) *For any two points ζ_1 and ζ_2 on S_0 there exists at least one γ which joins ζ_1 to ζ_2. We define the distance between ζ_1 and ζ_2 as the infimum of lengths $|\gamma|$ for all such curves γ. We assume that the distance between ζ_1 and ζ_2 vanishes if and only if the points coincide.*

(c) *The topology induced by this distance is equivalent to the original topology of S_0.*

(d) *Any region Δ bounded by a Jordan curve γ has a finite positive area $|\Delta|$.*

(e) *The length $|\gamma|$ and the area $|\Delta|$ are extended to be finitely additive measures on the finitely additive classes generated by $\{\gamma\}$ and $\{\Delta\}$, respectively.*

(f) *To each $\zeta \in S_0$ there corresponds a simply connected open neighborhood $N(\zeta)$ with an exterior point and a positive number $k(\zeta)$ such that for any Jordan curve $\gamma \subset N(\zeta)$ that is the boundary of a region $\Delta \subset N(\zeta)$, the inequality*

$$|\Delta| < k(\zeta)|\gamma| \tag{1}$$

is valid.

Property (f) shall be referred to as the *regularity* of the metric.

Lemma. *There exist two positive numbers d and k such that any Jordan curve γ with $|\gamma| < d$ on S_0 bounds a region Δ whose area is*

$$|\Delta| < k|\gamma|.$$

For the proof denote by $\delta > 0$ the shortest distance from a point ζ to the boundary of $N(\zeta)$ and let $V(\zeta)$ be a neighborhood of ζ contained in the $\delta/2$-neighborhood of ζ. Since S_0 is compact, there exists a finite covering $V(\zeta_1), \cdots, V(\zeta_m)$ of S_0. We denote by d the minimum of the corresponding numbers $\delta_1/2, \cdots, \delta_m/2$, and by k the maximum of the corresponding constants $k(\zeta_1), \cdots, k(\zeta_m)$. Every Jordan curve γ on S_0 with $|\gamma| < d$ intersects at least one $V(\zeta_i)$ and is consequently contained in $N(\zeta_i)$. By condition (f)

$$|\Delta| < k(\zeta_i)|\gamma| \leq k|\gamma|,$$

and the assertion is proved.

1B. Let a closed surface S_0 be decomposed into two parts Δ' and Δ'' by excluding a finite number of disjoint Jordan curves γ_ν. Denote by L the total length of these curves.

Lemma. *There exists a constant k depending only on S_0 such that*

$$\min(|\Delta'|, |\Delta''|) \le kL. \tag{2}$$

We first consider the case where every γ_ν has length less than the number d of Lemma 1A. Then γ_ν bounds a region Δ_ν such that

$$|\Delta_\nu| < k|\gamma_\nu|.$$

Let Ω be the complement of the union of the $\bar{\Delta}_\nu$ with respect to S_0. Since Ω is connected, it is obvious that Ω is contained either in Δ' or Δ''; suppose the former is the case. Then

$$|\Delta''| \le \sum |\Delta_\nu| < k \sum |\gamma_\nu| = kL.$$

Next assume there exists at least one γ_ν with $|\gamma_\nu| \ge d$. Then clearly $L \ge d$ and we obtain for $k = |S_0|/2d$

$$\min(|\Delta'|, |\Delta''|) \le \frac{|S_0|}{2} = kd \le kL.$$

1C. Consider a closed or open curve γ on S_0 and let ζ be a point on S_0. Suppose there exists a simply connected open neighborhood $N(\zeta)$ with exterior points such that for every Jordan curve γ_1 in $N(\zeta)$ the part of γ in the set in $N(\zeta)$ encircled by γ_1 has total length less than $\bar{k}(\zeta)|\gamma_1|$, with $\bar{k}(\zeta)$ a constant depending only on $N(\zeta)$. Then γ is said to be *regular*. Using the notations of Lemma 1B set $\gamma' = \gamma \cap \Delta'$ and $\gamma'' = \gamma \cap \Delta''$.

Lemma. *If γ is a regular simple curve, there exists a constant k depending only on S_0 and γ such that*

$$\min(|\gamma'|, |\gamma''|) \le kL. \tag{3}$$

The proof is similar to that of Lemma 1B.

1D. Now consider a compact bordered surface $\bar{S}_0$ with a given metric. Taking two copies of $\bar{S}_0$ and identifying the corresponding points on the borders we construct a closed surface $\tilde{S}_0$. We assume that the new metric on $\tilde{S}_0$ induced by the given metric on $\bar{S}_0$ satisfies the regularity condition of 1A. Note that every *cross-cut* on $\bar{S}_0$, i.e., simple curve with end points on the border, gives a Jordan curve on $\tilde{S}_0$. Accordingly Lemmas 1B and 1C hold also in the case where $\bar{S}_0$ is decomposed into two parts by some cycles (Jordan curves) and cross-cuts. In Lemma 1C we can take the curve γ to be a part of the border of $\bar{S}_0$ provided γ satisfies the regularity condition.

Let $\overline{S}_0$ be decomposed into two parts by finitely many disjoint cross-cuts with total length L on $\overline{S}_0$. Suppose that a curve γ on the border is decomposed into two parts γ' and γ''.

Lemma. *If γ is regular, then*

$$\min(|\gamma'|, |\gamma''|) \leq hL, \tag{4}$$

where h is a positive constant depending only on $\overline{S}_0$ and γ.

2. Covering of subregions

2A. Let S be a finite covering surface of a finite closed or bordered base surface S_0. By the *relative boundary* of S with respect to S_0 we mean the set of all boundary points whose projections are interior points of S_0. By a standard method we can lift the given metric of the base surface S_0 to the covering surface S. For example, to determine the length $|\gamma|$ of a curve γ on S we divide γ into arcs γ_ν, each contained in a triangle of the triangulation of S. The length $|\gamma_\nu|$ of γ_ν is, by definition, that of its projection on S_0, and the length $|\gamma|$ is $\sum |\gamma_\nu|$. The area of a region on S is defined analogously.

2B. To study covering properties of S above S_0 we denote by S_ν the set of points ζ_0 of S_0 above which there are at least ν interior points of S, the branch points counted with their multiplicities. Clearly S_ν is an open set which consists of finitely many regions of S_0. For convenience we call S_ν the νth sheet of S (strictly speaking, the projection of the νth sheet of S on S_0). Evidently

$$S_1 \supset \cdots \supset S_\nu \supset \cdots.$$

The largest number n among the ν with $S_\nu \neq \varnothing$ is called the *maximum number* of sheets of S. If a point ζ_0 of the base surface S_0 is contained in exactly ν regions $S_1, \cdots, S_\nu$, then ζ_0 is clearly covered by ν interior points of S. Consequently, if we denote by $|S|$ and $|S_\nu|$ the areas of S and S_ν respectively, then

$$|S| = \sum |S_\nu|. \tag{5}$$

Similarly, to the relative boundary of S there correspond relative boundaries of S_ν, $\nu = 1, \cdots, n$. Let γ be a side of a triangle in the triangulation of S_0 interior to S_0. It is counted once or twice as a part of the relative boundary of S_ν according as it is a side of one or two triangles belonging to S_ν. Under this convention, if above the side γ interior to S_0 the total number of sides is n_0 for each of the sheets $S_1, \cdots, S_n$, then there are n_0 sides belonging to the relative boundary of S. We denote by L and L_ν the lengths of the relative boundaries of S and S_ν and conclude that

$$L = L_1 + \cdots + L_n. \tag{6}$$

We call the ratio

$$M = \frac{|S|}{|S_0|} \tag{7}$$

the *mean sheet number* of S above S_0. For a region Δ on the base surface S_0 let $S(\Delta)$ be the part of S above Δ. We define the mean sheet number of $S(\Delta)$ above Δ by

$$M(\Delta) = \frac{|S(\Delta)|}{|\Delta|}. \tag{8}$$

For a curve γ in the interior or on the border of S_0 let $S(\gamma)$ be the part of S which lies above γ. Then the *mean sheet number* $M(\gamma)$ of $S(\gamma)$ above γ is, by definition,

$$M(\gamma) = \frac{|S(\gamma)|}{|\gamma|}. \tag{9}$$

2C. We are ready to state Ahlfors' [11] well-known covering theorem:

Theorem. *Any finite covering surface S of a finite surface S_0 satisfies the inequality*

$$|M - M(\Delta)| \leq \frac{k}{|\Delta|} L, \tag{10}$$

where k is a constant depending only on the metric of S_0.

Proof. For $S(\Delta)$ we have $|S(\Delta)| = \sum |S_\nu \cap \Delta|$ with $|S_\nu \cap \Delta| \leq |S_\nu|$ and

$$M_\nu(\Delta) = \frac{|S_\nu \cap \Delta|}{|\Delta|} \leq \frac{|S_\nu|}{|\Delta|}.$$

Clearly

$$M_\nu = \frac{|S_\nu|}{|S_0|} \leq \frac{|S_\nu|}{|\Delta|}.$$

Since M_ν and $M_\nu(\Delta)$ are nonnegative, we obtain

$$|M_\nu - M_\nu(\Delta)| \leq \frac{|S_\nu|}{|\Delta|}. \tag{11}$$

Next we replace S_ν by the set $S_0 - \bar{S}_\nu$. Clearly $\Delta - \overline{S_\nu \cap \Delta} \subset S_0 - \bar{S}_\nu$ and consequently

$$1 - M_\nu(\Delta) \leq \frac{|S_0| - |S_\nu|}{|\Delta|}.$$

Since

$$1 - M_\nu = \frac{|S_0| - |S_\nu|}{|S_0|} \leq \frac{|S_0| - |S_\nu|}{|\Delta|},$$

and the quantities $1-M_\nu(\Delta)$, $1-M_\nu$ are nonnegative, we conclude that

$$|M_\nu - M_\nu(\Delta)| \leq \frac{|S_0|-|S_\nu|}{|\Delta|}. \tag{12}$$

From (11) and (12) it follows that

$$|M_\nu - M_\nu(\Delta)| \leq \frac{1}{|\Delta|}\min(|S_\nu|, |S_0|-|S_\nu|). \tag{13}$$

By virtue of Lemma 1B this implies

$$|M_\nu - M_\nu(\Delta)| \leq \frac{k}{|\Delta|} L_\nu$$

with $\nu=1,\cdots,n$. On summing for all ν we obtain

$$|M-M(\Delta)| = \left|\sum (M_\nu - M_\nu(\Delta))\right| \leq \frac{k}{|\Delta|}\sum L_\nu = \frac{k}{|\Delta|} L.$$

Remark. If S is a covering surface which has no relative boundary above S_0, i.e., if $L=0$, then $M=M(\Delta)$ for every $\Delta\subset S_0$.

3. Covering of curves

3A. We add the following assumption to the requirements (a) to (f) in 1A:

(g) *Every curve which appears in the triangulation of S_0 and in its given subdivisions is regular.*

This assures the existence of sufficiently many regular curves on S_0. We next show that a similar covering theorem holds for the mean sheet number $M(\gamma)$ of S above a given *regular* curve γ on S_0. If S_0 is a bordered surface whose border consists of regular curves, then γ may lie on the border of S_0, with $S(\gamma)$ modified accordingly.

Theorem. *There exists a constant k depending only on γ and the metric of S_0 such that*

$$|M-M(\gamma)| \leq kL. \tag{14}$$

The proof will be furnished in 3B to 3D.

3B. For $S(\gamma)$ we have $|S(\gamma)| = \sum |S_\nu \cap \gamma|$.

First we discuss the case where γ decomposes S_0 into two parts. We denote by Δ one of the two regions whose area is not larger than that of the other. We note that $S_\nu \cap \Delta$ has a boundary of length at most $|S_\nu \cap \gamma| + L_\nu$, and by Lemma 1B we conclude that

$$|S_\nu \cap \Delta| \leq k(|S_\nu \cap \gamma| + L_\nu). \tag{15}$$

On replacing $S_\nu \cap \Delta$ by the set $\Delta - \overline{S_\nu \cap \Delta}$ we obtain

$$|\Delta| - |S_\nu \cap \Delta| \le k(|\gamma| - |S_\nu \cap \gamma| + L_\nu) \tag{16}$$

with the same constant k. Division of (15) and (16) by $|\Delta|$ gives

$$M_\nu(\Delta) \le \frac{k}{|\Delta|}(|S_\nu \cap \gamma| + L_\nu) \tag{17}$$

and

$$1 - M_\nu(\Delta) \le \frac{k}{|\Delta|}(|\gamma| - |S_\nu \cap \gamma| + L_\nu). \tag{18}$$

We observe that these inequalities continue to hold if $M_\nu(\Delta)$ is replaced by $M_\nu(\gamma) = |S_\nu \cap \gamma|/|\gamma|$. In fact by Lemma 1B, $1/|\gamma| \le k/|\Delta|$ and consequently

$$M_\nu(\gamma) = \frac{1}{|\gamma|}|S_\nu \cap \gamma| \le \frac{k}{|\Delta|}|S_\nu \cap \gamma| \le \frac{k}{|\Delta|}(|S_\nu \cap \gamma| + L_\nu).$$

Similarly

$$\begin{aligned} 1 - M_\nu(\gamma) = \frac{1}{|\gamma|}(|\gamma| - |S_\nu \cap \gamma|) &\le \frac{k}{|\Delta|}(|\gamma| - |S_\nu \cap \gamma|) \\ &\le \frac{k}{|\Delta|}(|\gamma| - |S_\nu \cap \gamma| + L_\nu). \end{aligned}$$

It follows that

$$|M_\nu(\Delta) - M_\nu(\gamma)| \le \frac{k}{|\Delta|}(|S_\nu \cap \gamma| + L_\nu) \tag{19}$$

and

$$|M_\nu(\Delta) - M_\nu(\gamma)| \le \frac{k}{|\Delta|}(|\gamma| - |S_\nu \cap \gamma| + L_\nu). \tag{20}$$

From these inequalities we deduce that

$$|M_\nu(\Delta) - M_\nu(\gamma)| \le \frac{k}{|\Delta|}[L_\nu + \min(|S_\nu \cap \gamma|, |\gamma| - |S_\nu \cap \gamma|)]. \tag{21}$$

By Lemma 1C (see also 1D) we have

$$\min(|S_\nu \cap \gamma|, |\gamma| - |S_\nu \cap \gamma|) \le k'' L_\nu$$

and therefore

$$|M_\nu(\Delta) - M_\nu(\gamma)| \le k' L_\nu, \tag{22}$$

where k' is a constant depending only on γ. Summing (22) for $\nu = 1, \cdots, n$ gives

$$|M(\Delta) - M(\gamma)| \le \sum |M_\nu(\Delta) - M_\nu(\gamma)| \le k' L. \tag{23}$$

By virtue of Theorem 2C we obtain

$$|M - M(\gamma)| \leq \bar{k}L. \tag{24}$$

3C. In the general case we deduce by using arguments analogous to those in 2C that

$$|M(\gamma) - M(\gamma')| \leq \frac{k}{|\gamma'|} L, \tag{25}$$

where γ' is an arc of any regular curve γ and k is a constant depending only on γ. For the sake of completeness we shall supply a proof.

Clearly $|S(\gamma')| = \sum |S_\nu \cap \gamma'|$ and $|S_\nu \cap \gamma'| \leq |S_\nu \cap \gamma|$. Hence

$$M_\nu(\gamma') = \frac{|S_\nu \cap \gamma'|}{|\gamma'|} \leq \frac{|S_\nu \cap \gamma|}{|\gamma'|}$$

and

$$M_\nu(\gamma) = \frac{|S_\nu \cap \gamma|}{|\gamma|} \leq \frac{|S_\nu \cap \gamma|}{|\gamma'|}.$$

Accordingly

$$|M_\nu(\gamma) - M_\nu(\gamma')| \leq \frac{|S_\nu \cap \gamma|}{|\gamma'|}. \tag{26}$$

Since $\gamma' \cap (S_0 - \bar{S}_\nu)$ is contained in $\gamma \cap (S_0 - \bar{S}_\nu)$,

$$|\gamma'| - |S_\nu \cap \gamma'| \leq |\gamma| - |S_\nu \cap \gamma|,$$

and a fortiori

$$1 - M_\nu(\gamma') \leq \frac{|\gamma| - |S_\nu \cap \gamma|}{|\gamma'|}.$$

On the other hand,

$$1 - M_\nu(\gamma) = \frac{|\gamma| - |S_\nu \cap \gamma|}{|\gamma|} \leq \frac{|\gamma| - |S_\nu \cap \gamma|}{|\gamma'|}$$

and we have

$$|M_\nu(\gamma) - M_\nu(\gamma')| \leq \frac{|\gamma| - |S_\nu \cap \gamma|}{|\gamma'|}. \tag{27}$$

From (26) and (27) it follows that

$$|M_\nu(\gamma) - M_\nu(\gamma')| \leq \frac{1}{|\gamma'|} \min\left(|S_\nu \cap \gamma|, |\gamma| - |S_\nu \cap \gamma|\right).$$

By virtue of Lemma 1C

$$|M_\nu(\gamma) - M_\nu(\gamma')| \leq \frac{k}{|\gamma'|} L_\nu, \tag{28}$$

where k is a constant depending only on γ.

On summing (28) for all ν we obtain (25).

3D. After these preliminaries let γ be a regular curve on S_0. Take an arbitrary point ζ_0 on the curve γ, a small simply connected open neighborhood $N(\zeta_0)$, and an arc γ' of γ in $N(\zeta_0)$. Form a regular Jordan curve γ^* in $N(\zeta_0)$ by adding a suitable arc to γ'. This is certainly possible by (g) of 3A. It is clear that γ^* decomposes S_0. For γ and γ' we have by (25)

$$|M(\gamma)-M(\gamma')| < \frac{k'}{|\gamma'|}L.$$

Similarly for γ^* and γ'

$$|M(\gamma^*)-M(\gamma')| < \frac{k^*}{|\gamma'|}L.$$

Furthermore, since γ^* decomposes S_0, on applying (24) we obtain

$$|M-M(\gamma^*)| < \bar{k}L.$$

From these three estimates we deduce the inequality

$$|M-M(\gamma)| < kL,$$

which we set out to prove.

§2. EULER CHARACTERISTIC

The classical Hurwitz formula gives a relation between the Euler characteristic of the base surface and that of the covering surface without relative boundary. Making use of mean sheet numbers discussed in §1 we can now give an extension of the Hurwitz formula to covering surfaces with relative boundaries. This is Ahlfors' main theorem on the Euler characteristic.

4. Preliminaries

4A. Consider a finite (closed or bordered) surface S with a given triangulation. Let V, E, and F be the numbers of interior vertices, interior edges, and faces of S, respectively. The Euler characteristic of S is, by definition,

$$e = e(S) = -V+E-F.$$

It is well known that e is independent of the triangulation and topologically invariant. If some interior vertices and edges are removed from the triangulation, the resulting subregions Ω_j have Euler characteristics $e(\Omega_j)$ in the original triangulation, with

$$e = \sum e(\Omega_j) - \bar{V} + \bar{E}. \tag{29}$$

Here $\bar{V}$ and $\bar{E}$ are the numbers of remaining vertices and edges, respectively.

We only consider the case where these vertices and edges form (disjoint) cross-cuts γ with end points on the border of S and cycles σ interior to S. The contribution to $-\bar{V}+\bar{E}$ of every γ is 1 and of every σ is 0:

$$e = \sum e(\Omega_j)+n(\gamma), \tag{30}$$

where $n(\gamma)$ denotes the number of cross-cuts. From this it is easy to see that if S is simply connected, then $e=-1$; if S is a planar surface with q contours, then $e=q-2$; if S has q contours and genus g, then $e=q+2g-2$.

On setting $e^+=\max(e, 0)$ and on denoting by $N_1(\Omega_j)$ the number of simply connected regions Ω_j, we obtain the following equivalent formulation of (30):

$$e = \sum e^+(\Omega_j)+n(\gamma)-N_1(\Omega_j). \tag{31}$$

4B. Now let S be a finite covering surface of a base surface S_0. Our task is to study the relation between the Euler characteristics of S and S_0.

To this end let S_ν, $\nu=1, \cdots, n$, be the νth sheet of S. If V_ν, E_ν, and F_ν are the numbers of interior vertices, interior edges, and faces of S_ν, respectively, then $e_\nu=-V_\nu+E_\nu-F_\nu$ is the sum of the Euler characteristics of the regions which form the sheet S_ν. For edges and faces we obviously have $E=\sum E_\nu$, $F=\sum F_\nu$. In contrast, the vertices satisfy the inequality $V\leq\sum V_\nu$ because we count interior vertices of S without regard to their multiplicities. In terms of the sum v of the orders of the branch points of S we have $V=\sum V_\nu-v$ and therefore

$$e = \sum e_\nu+v. \tag{32}$$

If S has no relative boundary above S_0, then every e_ν is equal to the characteristic e_0 of the base surface S_0, and we have the classical Hurwitz formula:

$$e = ne_0+v, \tag{33}$$

where n is the sheet number of S. Thus (32) can be regarded as a purely topological extension of the Hurwitz formula (33). It is difficult to estimate e_ν in the case where the covering behavior of S above S_0 is complicated. Ahlfors' main theorem will give a *metric-topological extension of the Hurwitz formula* in the general case.

5. Cross-cuts and regions

5A. For preparation we first consider the case of a planar base surface S_0.

Suppose that the border of S_0 consists of $q>2$ contours, some of which may reduce to points. We divide the base surface S_0 into two simply connected subsurfaces S_0' and S_0'' by removing q disjoint regular cross-cuts $\beta_1, \cdots, \beta_q$, each of finite length. These cross-cuts are fixed through the proof. We may assume that they consist of interior edges in the triangulation of S_0 or in some subdivision of it. Furthermore, we choose $\beta_1, \cdots, \beta_q$ so that the covering surface S possesses no branch points above them. We also assume that the triangulation of S_0 is lifted to S and consider on S all interior edges above $\beta_1, \cdots, \beta_q$. Evidently they form a system of disjoint cross-cuts γ on S. If we exclude them, S is decomposed into a finite number $N(\Omega)$ of regions Ω and we have as a direct consequence of (31)

$$e \geq n(\gamma) - N(\Omega). \tag{34}$$

5B. To deduce the main theorem from inequality (34) we introduce a classification of cross-cuts γ and regions Ω.

First we separate classes of regions $\{\Omega_1\}, \cdots, \{\Omega_p\}$ with $p \leq n(\gamma)$ and classes of cross-cuts $\{\gamma_1\}, \cdots, \{\gamma_p\}$ from other regions Ω and cross-cuts γ. Assume that there exists a region Ω having only one cross-cut on its boundary. The totality of such regions forms the first class $\{\Omega_1\}$, and the totality of the corresponding cross-cuts forms the first class $\{\gamma_1\}$. The class $\{\Omega_2\}$ consists of all regions with only one cross-cut on their boundaries not belonging to the class $\{\gamma_1\}$, and the corresponding cross-cuts form the class $\{\gamma_2\}$. Inductively, $\{\Omega_\nu\}$ is formed by regions with only one cross-cut on their boundaries not belonging to $\{\gamma_1\} \cup \cdots \cup \{\gamma_{\nu-1}\}$, and the corresponding cross-cuts constitute the class $\{\gamma_\nu\}$. We continue the process until all regions have been separated or each remaining region has no new cross-cuts or has at least two new cross-cuts on its boundary.

We observe that each region Ω_ν has a cross-cut γ_ν on its boundary.

Conversely, except for a special case, each cross-cut γ_ν belongs to a unique region, say Ω_ν. For the proof we shall show that every cross-cut γ_ν decomposes S into two parts and that all cross-cuts γ belonging to the part containing the region Ω_ν bounded by γ_ν belong to $\{\gamma_1\} \cup \cdots \cup \{\gamma_{\nu-1}\}$. In the case $\nu = 1$ the truth of the assertion is evident. We assume that it holds for subindices less than ν. The region Ω_ν has the cross-cut γ_ν and some cross-cuts belonging to a class lower than $\{\gamma_\nu\}$ on its boundary. By assumption each γ_{ν_1} with $\nu_1 < \nu$ decomposes S into two regions. The region which does not contain Ω_ν contains only cross-cuts belonging to classes lower than $\{\gamma_{\nu_1}\}$. Consequently the cross-cuts γ_ν decompose S into two parts and the set of points which can be joined to a point of Ω_ν by continuous curves without intersecting the cross-cut γ_ν consists of the region Ω_ν and some regions belonging to classes lower than $\{\Omega_\nu\}$. Since this set contains only cross-cuts belonging to $\{\gamma_1\} \cup \cdots \cup \{\gamma_{\nu-1}\}$, our assertion follows.

Consider the case where a single cross-cut γ_ν belongs to the boundaries of two different regions Ω_ν. By the fact just proved all other cross-cuts γ must belong to a class lower than $\{\gamma_\nu\}$. Accordingly, in the present case all cross-cuts γ belong to the separated classes $\{\gamma_1\}, \{\gamma_2\}, \cdots, \{\gamma_p\}$, and γ_ν is a unique cross-cut of the highest class $\{\gamma_p\}$. For convenience we shall refer to this exceptional situation as *case A*. Except for this case there exists a one-to-one correspondence between the separated regions and the separated cross-cuts.

As for a remaining unseparated region Ω it is easy to see that its boundary contains

(a) separated cross-cuts only, or

(b) at least two unseparated cross-cuts.

We note that alternative (a) is rather exceptional because if a region Ω is bounded by only separated cross-cuts γ_ν, then each γ_ν decomposes S into two parts and the part not containing Ω contains only cross-cuts belonging to a class lower than $\{\gamma_\nu\}$. In this situation, to be called *case B*, all cross-cuts γ must belong to the separated classes. Evidently there exists at most one region Ω satisfying (a).

5C. If there are unseparated regions, we continue their classification. If Ω does not belong to any class $\{\Omega_\nu\}$, $\nu=1, \cdots, p$, then its boundary contains at least two unseparated cross-cuts. According as the number of unseparated cross-cuts is less than q or at least q, we let the unseparated region Ω belong to the class $\{\Omega'\}$ or $\{\Omega''\}$. Furthermore, according as the unseparated cross-cut γ bounds two regions of class $\{\Omega'\}$, or one region of this class and another region of $\{\Omega''\}$, or two regions of the class $\{\Omega''\}$, we denote it by γ_{11} or γ_{12} or γ_{22}, respectively. Our classification is herewith complete. The scheme is

$$\begin{array}{ccc} \{\Omega_1\}\cdots\{\Omega_p\} & \{\Omega'\} & \{\Omega''\} \\ \{\gamma_1\}\cdots\{\gamma_p\} & \overbrace{\{\gamma_{11}\}\,\{\gamma_{12}\}} & \overbrace{\{\gamma_{12}\}\,\{\gamma_{22}\}} \end{array}$$

and the main relations are:

(a) Ω_ν and γ_ν correspond to each other bijectively,

(b) every Ω' contains at least 2 and at most $q-1$ cross-cuts γ_{11} or γ_{12} on its boundary,

(c) Ω'' contains at least q cross-cuts γ_{12} and γ_{22} on its boundary.

6. Main theorem on Euler characteristic

6A. At this point we shall state Ahlfors' extension of Hurwitz's formula to covering surfaces with relative boundaries.

Consider a finite (closed or bordered) surface S_0 and a finite covering surface S of S_0. Let e_0 and e be the Euler characteristics of S_0 and S, and set $e^+ = \max(e, 0)$. Denote by M the mean sheet number of S, and by L the total length of the relative boundary of S.

Theorem. *There exists a constant k depending on S_0 but independent of S such that*

$$e^+ \geq Me_0 - kL. \tag{35}$$

The proof will be given in 6B to 7C.

6B. We start with inequality (34). We consider the case where the cross-cuts γ_ν do not exhaust the set of all cross-cuts γ; in other words, we exclude cases A and B. Then by (a) of 5C inequality (34) can be written in the form

$$e \geq [n(\gamma_{11}) + \tfrac{1}{2}n(\gamma_{12}) - N(\Omega')] + [n(\gamma_{22}) + \tfrac{1}{2}n(\gamma_{12}) - N(\Omega'')]. \tag{36}$$

Let Ω_i' be any Ω'. Its boundary contains $n_i(\gamma_{11})$ cross-cuts γ_{11} and $n_i(\gamma_{12})$ cross-cuts γ_{12}. According to (b) of 5C, $n_i(\gamma_{11}) + n_i(\gamma_{12}) \geq 2$. On summing for every Ω' we have

$$\sum n_i(\gamma_{11}) + \sum n_i(\gamma_{12}) \geq 2N(\Omega').$$

However,

$$\sum n_i(\gamma_{11}) = 2n(\gamma_{11}), \qquad \sum n_i(\gamma_{12}) = n(\gamma_{12}),$$

and therefore

$$n(\gamma_{11}) + \tfrac{1}{2}n(\gamma_{12}) - N(\Omega') \geq 0.$$

Similarly we obtain in view of (c) of 5C

$$n(\gamma_{22}) + \tfrac{1}{2}n(\gamma_{12}) \geq \frac{q}{2} N(\Omega'').$$

Hence

$$e \geq n(\gamma_{22}) + \tfrac{1}{2}n(\gamma_{12}) - N(\Omega'') \geq \frac{q-2}{q} (n(\gamma_{22}) + \tfrac{1}{2}n(\gamma_{12}))$$

and we conclude that

$$e \geq \frac{q-2}{q} n(\gamma_{22}). \tag{37}$$

6C. We shall estimate $n(\gamma_{22})$ from below. To this end we recall that each cross-cut γ is above one of the curves $\beta_1, \cdots, \beta_q$ used to decompose the base surface S_0 into two simply connected regions S_0' and S_0''. We denote by $\lambda(\gamma)$ the length of the cross-cut γ above β divided by the length of β:

$$\lambda(\gamma) = \frac{|\gamma|}{|\beta|}. \tag{38}$$

The sum of the mean sheet numbers of S above $\beta_1, \cdots, \beta_q$ is

$$\sum M(\beta_i) = \sum \lambda(\gamma_\nu) + \sum \lambda(\gamma_{11}) + \sum \lambda(\gamma_{12}) + \sum \lambda(\gamma_{22}).$$

By virtue of Theorem 3A there exists a constant k depending only on the metric of S_0 and on the choice of the arcs $\beta_1, \cdots, \beta_q$ such that

$$\sum M(\beta_i) \geq qM - kL.$$

It follows that

$$\sum \lambda(\gamma_{22}) \geq qM - kL - (\sum \lambda(\gamma_\nu) + \sum \lambda(\gamma_{11}) + \sum \lambda(\gamma_{12})).$$

Here we observe that for each γ clearly $\lambda(\gamma) \leq 1$ and, in particular, $\sum \lambda(\gamma_{22}) \leq n(\gamma_{22})$ and $(q-2)/q < 1$. Thus we obtain from (37)

$$e \geq (q-2)M - kL - (\sum \lambda(\gamma_\nu) + \sum \lambda(\gamma_{11}) + \sum \lambda(\gamma_{12})). \tag{39}$$

6D. We wish to show that each of the three sums on the right-hand side of (39) is at most kL. To this end we apply the covering theorems. We decomposed S_0 into two simply connected regions S_0' and S_0'' by $\beta_1, \cdots, \beta_q$. Now we take S_0' and S_0'' as base surfaces. Let Ω be a covering surface of S_0' or S_0'', and let $M_\Omega(\beta_\nu)$ be the mean sheet number of Ω above β_ν. Then by Theorem 3A we can find a constant k such that for every pair of curves β_μ and β_ν

$$|M_\Omega(\beta_\mu) - M_\Omega(\beta_\nu)| \leq \frac{q-2}{q-1} kL(\Omega), \tag{40}$$

where $L(\Omega)$ denotes the length of the relative boundary of Ω above S_0' or S_0''.

Choose a separated region Ω_ν as Ω. Clearly Ω_ν is a covering surface of S_0' or S_0''. By (a) of 5C a unique γ_ν corresponds to each Ω_ν. We may assume that γ_ν is above β_1. Then for $i=2, \cdots, q$ we have by (40)

$$\lambda(\gamma_\nu) \leq M_{\Omega_\nu}(\beta_1) \leq M_{\Omega_\nu}(\beta_i) + \frac{q-2}{q-1} kL(\Omega_\nu).$$

On summing for $i=2, \cdots, q$ we obtain

$$(q-1)\lambda(\gamma_\nu) \leq \sum M_{\Omega_\nu}(\beta_i) + (q-2)kL(\Omega_\nu)$$

and on adding $\lambda(\gamma_\nu) \leq M_{\Omega_\nu}(\beta_1)$,

$$q\lambda(\gamma_\nu) \leq \sum_{\Omega_\nu} \lambda(\gamma) + (q-2)kL(\Omega_\nu), \tag{41}$$

because the total sum $\sum M_{\Omega_\nu}(\beta_i)$ of the mean sheet numbers of Ω_ν is equal to the sum $\sum \lambda(\gamma)$ for all γ of Ω_ν.

6E. We sum (41) for all regions belonging to $\{\Omega_1\} \cup \cdots \cup \{\Omega_p\}$. The sum of the first terms on the right-hand side of (41) is at most $2 \sum \lambda(\gamma_\nu)$, because each cross-cut on the boundary of Ω_ν is contained in the boundaries of at most *two* separated regions. The sum of the lengths of the relative boundaries of Ω_ν with respect to S_0' or S_0'' is at most the length L of the relative boundary of S with respect to S_0. Consequently the sum of the remaining terms is at most $(q-2)kL$. The sum $q\lambda_\alpha$ of the $q\lambda(\gamma_\nu)$ for all Ω_ν is precisely the total sum $q \sum \lambda(\gamma_\nu)$ because we excluded case A; in case A, $\lambda(\gamma_p)$ is counted twice in λ_α, where γ_p is a unique cross-cut belonging to the highest class $\{\gamma_p\}$ and therefore $\lambda_\alpha \geq \sum \lambda(\gamma_\nu)$. We infer that

$$(42) \qquad q \sum \lambda(\gamma_\nu) \leq 2 \sum \lambda(\gamma_\nu) + (q-2)kL,$$

and a fortiori

$$(43) \qquad \sum \lambda(\gamma_\nu) \leq kL.$$

This inequality holds also in case A.

6F. It remains to estimate $\sum \lambda(\gamma_{11})$ and $\sum \lambda(\gamma_{12})$ in (39). A region $\Omega' \in \{\Omega'\}$ possesses at most $q-1$ cross-cuts γ_{11} or γ_{12} on its boundary. Therefore, for each Ω' there exists at least one arc β_i above which there is no cross-cut of $\{\gamma_{11}\}$ and $\{\gamma_{12}\}$. If $M_{\Omega'}(\beta_i)$ denotes the mean sheet number of Ω' above β_i, then

$$\sum_{\Omega'} \lambda(\gamma_{11}) + \sum_{\Omega'} \lambda(\gamma_{12}) \leq \sum_{j \neq i} M_{\Omega'}(\beta_j) \leq (q-1)\, M_{\Omega'}(\beta_i) + (q-2)\, kL(\Omega'),$$

where on the left the sums are taken over all γ_{11} and γ_{12} relative to Ω'.

On summing these inequalities for all Ω' the left-hand side becomes $2 \sum \lambda(\gamma_{11}) + \sum \lambda(\gamma_{12})$. The sum $\sum M_{\Omega'}(\beta_i)$ on the right depends only on separated cross-cuts of $\{\gamma_\nu\}$, as is seen from the choice of β_i. Since each cross-cut γ_ν belongs to the boundary of at most one Ω', we have

$$\sum M_{\Omega'}(\beta_i) \leq \sum \lambda(\gamma_\nu).$$

Clearly $\sum L(\Omega') \leq L$. By virtue of inequality (43) we obtain

$$\begin{aligned} 2 \sum \lambda(\gamma_{11}) + \sum \lambda(\gamma_{12}) &\leq (q-1) \sum \lambda(\gamma_\nu) + (q-2)kL \\ &\leq (2q-3)kL \leq k'L. \end{aligned}$$

The main theorem for the covering surface S of a planar base surface S_0 is herewith proved except for cases A and B.

In cases A and B inequality (43) continues to hold. Since $\sum M_{\Omega'}(\beta_i) = \sum \lambda(\gamma_\nu)$, we have

$$(44) \qquad qM \leq \sum M_{\Omega'}(\beta_i) + kL = \sum \lambda(\gamma_\nu) + kL \leq k'L.$$

Thus the mean sheet number M of S is at most $k''L$, and for this reason the main theorem holds if we take k to be sufficiently large in (35). The proof of the main theorem is complete for a planar S_0.

7. Extension to positive genus

7A. We turn to the case where the base surface S_0 is not planar. It is well known that we can transform S_0 into a planar surface $\tilde{S}_0$ with the same Euler characteristic e_0 by excluding a set of disjoint cycles $\tilde{\beta}$ from S_0. On the covering surface S consider the curves above the cycles $\tilde{\beta}$. In general these curves consist of some cross-cuts $\tilde{\gamma}$ and some cycles $\tilde{\sigma}$ on S; if we exclude them from S, we can decompose S into some subsurfaces $\tilde{S}$.

Each $\tilde{S}$ is clearly a covering surface of the base surface $\tilde{S}_0$. Let $n(\tilde{\gamma})$ be the number of cross-cuts $\tilde{\gamma}$. Then by (31)

$$e = \sum e^+(\tilde{S}) + n(\tilde{\gamma}) - N_1(\tilde{S}),$$

where $N_1(\tilde{S})$ is the number of simply connected subsurfaces $\tilde{S}$.

7B. By a well-known theorem in topology a surface is decomposed into at most two parts by excluding a cross-cut; if the two parts are simply connected, then the original surface is simply connected. Applying this theorem repeatedly we see that if we decompose a given surface by excluding only cross-cuts, the number of subsurfaces is greater than that of the cross-cuts by at most one. In particular, if the given surface is not simply connected, then the number of simply connected subsurfaces is at most that of the cross-cuts.

We apply this result to the above decomposition. First decompose S into some regions Ω by excluding $n(\tilde{\gamma})$ cross-cuts. If $e \geq 0$, the number of simply connected regions Ω is at most $n(\tilde{\gamma})$. Next exclude the cycles $\tilde{\sigma}$ from these regions Ω to obtain the original decomposition $S = (\bigcup \tilde{S}) \cup (\bigcup \tilde{\gamma}) \cup (\bigcup \tilde{\sigma})$. We cannot obtain new simply connected regions by excluding the cycles $\tilde{\sigma}$ from the regions Ω, because each region with a cycle on its boundary possesses at least one other contour. If $e \geq 0$, we thus have $N_1(\tilde{S}) \leq n(\tilde{\gamma})$ and $e^+ \geq \sum e^+(\tilde{S})$. On the other hand, if $e = -1$, then $N_1(\tilde{S}) = n(\tilde{\gamma}) + 1$, and therefore

$$e = \sum e^+(\tilde{S}) + n(\tilde{\gamma}) - N_1(\tilde{S}) = \sum e^+(\tilde{S}) - 1,$$

whence it follows that all $e^+(\tilde{S}) = 0$. In both cases we thus obtain

$$e^+ \geq \sum e^+(\tilde{S}). \tag{45}$$

7C. In the planar case we have by (35) for each $\tilde{S}$

$$e^+(\tilde{S}) \geq e_0 M(\tilde{S}) - kL(\tilde{S}),$$

where $M(\tilde{S})$ denotes the mean sheet number of $\tilde{S}$ above $\tilde{S}_0$ and $L(\tilde{S})$ is the length of the relative boundary of $\tilde{S}$ above $\tilde{S}_0$. By virtue of (45) we obtain

$$e^+ \geq e_0 \sum M(\tilde{S}) - k \sum L(\tilde{S}). \tag{46}$$

Since

$$M = \sum M(\tilde{S}) \quad \text{and} \quad L \geq \sum L(\tilde{S}),$$

we conclude that

$$e^+ \geq e_0 M - kL.$$

This completes the proof of the main theorem.

§3. ISLANDS AND PENINSULAS

The components of the set covering a region Δ of the base surface are classified as islands or peninsulas according as they do not or do have a relative boundary point of the covering surface on their boundaries. The mean sheet numbers of islands and peninsulas are the counterparts of the counting function and the proximity function. In terms of these numbers we shall establish the fundamental inequality for arbitrary covering surfaces. This inequality is the analogue of the second main theorem I.5C.

8. Fundamental inequality

8A. Let $\hat{S}$ be an open or, equivalently, infinite covering surface of a finite base surface S_0. Consider finite covering surfaces $S \subset \hat{S}$ of S_0. The surface $\hat{S}$ is called *regularly exhaustible* if

$$\liminf_{S \to \hat{S}} \frac{L(S)}{M(S)} = 0, \tag{47}$$

where $L(S)$ is the length of the relative boundary and $M(S)$ the mean sheet number of S. In this case every subexhaustion $S \to \hat{S}$ with

$$\lim_{S \to \hat{S}} \frac{L(S)}{M(S)} = 0$$

is called *regular*. We shall later (in 12A and 18A) give explicit tests for regular exhaustibility. This property is necessary in order that the remainder $O(L)$ in the theorems to be established be negligible.

By Theorem 2C we have for a region Δ on S_0

$$|M - M(\Delta)| \leq kL.$$

Here k is a constant depending only on Δ and the metric of S_0.

We consider a component D of the part of S above Δ. If D has no relative boundary point of S on its boundary, it is called an *island*, otherwise a *peninsula*. The (mean) sheet number of an island gives an integer value to $M(\Delta)$. If we denote by $n(\Delta)$ the sum of the sheet numbers of all islands above Δ, and by $\mu(\Delta)$ the part of $M(\Delta)$ contributed by the peninsulas above Δ, we have the following counterpart of the first main theorem I.2E (II.9A, III.1G):

Theorem. *The islands and peninsulas above Δ satisfy the equation*

$$n(\Delta)+\mu(\Delta) = M+O(L), \tag{48}$$

where $O(L)$ depends only on Δ.

If $\hat{S}$ is regularly exhaustible, then

$$\liminf_{S\to\hat{S}} \frac{n(\Delta)}{M} \leq 1. \tag{49}$$

8B. Let g be the genus and α the number of contours of S_0. The Euler characteristic e_0 of S_0 is

$$e_0 = e(S_0) = 2g+\alpha-2. \tag{50}$$

Consider $q \geq 2$ disjoint simply connected regions $\Delta_1, \cdots, \Delta_q$ on S_0. Denote by $b(\Delta_\nu)$ the sum of the orders of branch points of all islands above Δ_ν. We shall establish the following extension of Ahlfors' [11] *fundamental inequality*:

Theorem. *For an arbitrary covering surface $\hat{S}$ of a closed Riemann surface S_0, and a finite covering surface $S \subset \hat{S}$ of S_0*

$$(e_0+q)M < \sum n(\Delta_\nu)-\sum b(\Delta_\nu)+e^+(S)+O(L). \tag{51}$$

The theorem shows that there can exist only relatively few regions Δ_ν sparsely covered with islands. This is a striking metric-topological analogue of Theorem I.5C.

In the form (51) the theorem was established by Sario [7] and Noshiro. For important earlier work we refer to Ahlfors [11], [12], Dufresnoy [11], Kunugui [3], Noshiro [4], and Tumura [4].

9. Auxiliary estimates

9A. To prove (51) for $g \geq 1$ we cut S_0 into a planar surface $\tilde{S}_0$ with the same Euler characteristic e_0 by excluding from S_0 a set of disjoint cycles $\beta_1, \cdots, \beta_g$ which do not intersect any Δ_ν, $\nu = 1, \cdots, q$. From the covering surface S we exclude all curves above all cycles $\beta_1, \cdots, \beta_g$. Then as in 7A,

S is decomposed into disjoint covering surfaces $\tilde{S}$ of the base surface $\tilde{S}_0$. We know that

$$e^+(S) \geq \sum e^+(\tilde{S}_0). \tag{52}$$

9B. Consider the components D of the parts $\tilde{S}$ above $\Delta_1, \cdots, \Delta_q$. We shall denote islands by D_i and peninsulas by D_p. First we exclude all peninsulas D_p from $\tilde{S}$ and consider the remaining regions $\tilde{S}'$. If $\tilde{S}$ is simply connected, then so is every $\tilde{S}'$. In the case of a multiply connected $\tilde{S}$, some $\tilde{S}'$ may also be multiply connected.

Lemma. *The inequality*

$$\sum e^+(\tilde{S}') \leq e^+(\tilde{S}) \tag{53}$$

holds for arbitrary finite covering surfaces of finite base surfaces.

Proof. For a simply connected $\tilde{S}$ the statement is trivial, and we shall suppose that $\tilde{S}$ is *multiply connected*. Decompose $\tilde{S}$ into regions $\tilde{S}'$ and some peninsulas D_p by excluding $n(\gamma)$ cross-cuts γ and some cycles σ. We carry out this process in *two steps*. First we decompose $\tilde{S}$ into some regions G by excluding only $n(\gamma)$ cross-cuts γ. Then we cut these regions G by excluding the cycles σ.

We denote G by G_p or G' according as it contains at least one peninsula D_p or not. Clearly G' is identical with some $\tilde{S}'$. If G_p is simply connected, then since G_p contains a unique peninsula D_p, the boundary of G_p is contained in the boundary of D_p. As a consequence, by excluding some cycles σ from the region G_p we obtain only simply connected regions $\tilde{S}'$.

By (30) we have

$$e(\tilde{S}) = n(\gamma) + \sum e(G_p) + \sum e(G').$$

In terms of the numbers $N_1(G_p)$ and $N_1(G')$ of simply connected regions G_p and G', respectively, this can be written

$$e(\tilde{S}) = [n(\gamma) - N_1(G_p) - N_1(G')] + \sum_2 e(G_p) + \sum_2 e(G'), \tag{54}$$

where $\sum_2$ means the sum extended over multiply connected regions. Since $\tilde{S}$ is multiply connected by assumption, we obtain at most $n(\gamma)$ simply connected regions by excluding $n(\gamma)$ disjoint cross-cuts. For this reason the expression in brackets on the right-hand side of (54) is nonnegative, and we obtain

$$e(\tilde{S}) \geq \sum_2 e(G_p) + \sum_2 e(G'). \tag{55}$$

9C. Every multiply connected G_p is decomposed into peninsulas and regions $\tilde{S}'$ by excluding some cycles σ. We omit every cycle σ which bounds a simply connected $\tilde{S}'$ and a peninsula D_p. Then the remaining cycles σ decompose the region G_p into regions G'_p and multiply connected regions $\tilde{S}'$. However, every G'_p is multiply connected, because every remaining cycle σ separates the boundary components of G_p. Consequently, for each multiply connected G_p we have

$$e(G_p) = \sum e(G'_p) + \sum_2 e^+(\tilde{S}').$$

From this and from (55) we obtain the lemma.

10. Proof of the fundamental inequality

10A. We now exclude all islands D_i from these $\tilde{S}'$. Then each of the remaining regions $\tilde{\tilde{S}}$ is a covering surface of the base surface $\tilde{\tilde{S}}_0$ obtained by excluding the closure of the union of $\Delta_1, \cdots, \Delta_q$ from $\tilde{S}_0$. Since $\tilde{S}'$ is decomposed into the islands D_i and regions $\tilde{\tilde{S}}$ by excluding only cycles, we have

$$\sum e(\tilde{S}') = \sum e(\tilde{\tilde{S}}) + \sum e(D_i).$$

It follows that

$$\begin{aligned} -\sum e(D_i) &= \sum e(\tilde{\tilde{S}}) - \sum e(\tilde{S}') \\ &= \sum e^+(\tilde{\tilde{S}}) - N_1(\tilde{\tilde{S}}) - (\sum e^+(\tilde{S}') - N_1(\tilde{S}')) \\ &= \sum e^+(\tilde{\tilde{S}}) - \sum e^+(\tilde{S}') + N_1(\tilde{S}') - N_1(\tilde{\tilde{S}}), \end{aligned}$$

where $N_1(\tilde{S}')$ and $N_1(\tilde{\tilde{S}})$ are the numbers of simply connected regions $\tilde{S}'$ and $\tilde{\tilde{S}}$. We note that if $\tilde{S}'$ contains at least one island D_i, then we obtain only multiply connected regions $\tilde{\tilde{S}}$ by excluding all the islands D_i from $\tilde{S}'$. For this reason every simply connected $\tilde{\tilde{S}}$ is identical with some $\tilde{S}'$, and we conclude that $N_1(\tilde{S}') - N_1(\tilde{\tilde{S}}) \geq 0$.

We obtain

$$-\sum e(D_i) \geq \sum e^+(\tilde{\tilde{S}}) - \sum e^+(\tilde{S}'). \tag{56}$$

From this and from (53) it follows that

$$-\sum e(D_i) \geq \sum e^+(\tilde{\tilde{S}}) - e^+(\tilde{S}). \tag{57}$$

By Ahlfors' main theorem we have

$$-\sum e(D_i) \geq e(\tilde{\tilde{S}}_0) \sum M(\tilde{\tilde{S}}) - e^+(\tilde{S}) - \bar{k} \sum L(\tilde{\tilde{S}}), \tag{58}$$

where $M(\tilde{\tilde{S}})$ denotes the mean sheet number of $\tilde{\tilde{S}}$ above $\tilde{\tilde{S}}_0$, and $L(\tilde{\tilde{S}})$ is the length of the relative boundary of $\tilde{\tilde{S}}$. It is clear that $\sum L(\tilde{\tilde{S}})$ is majorized by the length $L(\tilde{S})$ of the relative boundary $\tilde{S}$ above $\tilde{S}_0$. By virtue of (10) we can replace $\sum M(\tilde{\tilde{S}})$ by $M(\tilde{S})$. We infer that

$$-\sum e(D_i) \geq e(\tilde{\tilde{S}}_0)M(\tilde{S}) - e^+(\tilde{S}) - kL(\tilde{S}), \tag{59}$$

where

$$e(\tilde{\tilde{S}}_0) = e_0 + q = 2g + \alpha - 2 + q.$$

10B. We recall that the covering surface S of the base surface S_0 is decomposed into regions $\tilde{S}$. We have obtained inequality (59) for each $\tilde{S}$. Now we sum these inequalities for all regions $\tilde{S}$ and obtain

$$-\sum e(D_i) \geq (e_0+q)\sum M(\tilde{S}) - k\sum L(\tilde{S}) - \sum e^+(\tilde{S}).$$

Here

$$\sum M(\tilde{S}) = M = M(S) \quad \text{and} \quad L \geq \sum L(\tilde{S}).$$

By inequality (52) we have

$$-\sum e(D_i) \geq (e_0+q)M - kL - e^+(S). \tag{60}$$

10C. An application of the Hurwitz formula (33) to each island D_i above every Δ_ν, $\nu = 1, \cdots, q$, gives

$$-\sum e(D_i) = \sum n(\Delta_\nu) - \sum b(\Delta_\nu), \tag{61}$$

where $n(\Delta_\nu)$ denotes the sum of sheet numbers of all islands of S, and $b(\Delta_\nu)$ is the sum of the orders of the branch points of all islands of S above Δ_ν. From (60) and (61) we obtain

$$\sum n(\Delta_\nu) - \sum b(\Delta_\nu) \geq (e_0+q)M - e^+(S) - kL. \tag{62}$$

The proof of the fundamental inequality (51) is herewith complete.

Using formula (48) we can rewrite (62) in the form

$$\sum \mu(\Delta_\nu) \leq -e_0 M - \sum b(\Delta_\nu) + e^+(S) + kL. \tag{63}$$

10D. It is important to note that in (62) and (63) we can replace the regions Δ_ν by q given points $a_\nu \in S$. Choose q disjoint simply connected regions Δ_ν with $a_\nu \in \Delta_\nu$. Obviously the number $\bar{n}(a_\nu)$ of interior points of S above a_ν counted by using *simple multiplicities* is at least $-\sum e(D_i)$ for the union of all islands above Δ_ν. It follows from (60) that

$$\sum \bar{n}(a_\nu) \geq (e_0+q)M - e^+(S) - kL. \tag{60$'$}$$

Thus we obtain

$$(e_0+q)M \leq \sum n(a_\nu) - \sum b(a_\nu) + e^+(S) + kL \tag{62$'$}$$

and

$$(63)' \qquad \sum (M - n(a_\nu)) \leq -e_0 M - \sum b(a_\nu) + e^+(S) + kL,$$

where $n(a_\nu)$ is the number of interior points of S above a_ν counted with their multiplicities, and $b(a_\nu)$ is the sum of the orders of the branch points of S above a_ν.

11. Defects and ramifications

11A. Let $\hat{S}$ be a *regularly exhaustible* covering surface with the exhaustion $S \to \hat{S}$ satisfying condition (47). We introduce the *defect*

$$(64) \qquad \gamma(\Delta_\nu) = \liminf \frac{\mu(\Delta_\nu)}{M(S)} = 1 - \limsup \frac{n(\Delta_\nu)}{M(S)},$$

the *ramification index*

$$(65) \qquad \varepsilon(\Delta_\nu) = \liminf \frac{b(\Delta_\nu)}{M(S)},$$

and the *Euler index*

$$\xi = \limsup \frac{e^+(S)}{M(S)},$$

all limits being directed ones for $S \to \hat{S}$.

We have arrived at the following metric-topological counterpart of the defect and ramification relation I.(39):

Theorem. *For an arbitrary regularly exhaustible covering surface of a finite surface*

$$(66) \qquad \sum \gamma(\Delta_\nu) + \sum \varepsilon(\Delta_\nu) \leq \xi - e_0.$$

11B. We proceed to study ramification properties more closely. Generalizing a result of Ahlfors [11] we shall give an explicit relation in terms of sheet numbers in ramified coverings. If every simply connected island D_i of $\hat{S}$ above Δ_ν has at least m_ν sheets, then $\hat{S}$ is said to be at least *m_ν-ply ramified* above Δ_ν.

Theorem. *The inequality*

$$(67) \qquad \sum \left(1 - \frac{1}{m_\nu}\right) \leq \xi - e_0$$

holds for regularly exhaustible covering surfaces $\hat{S}$ that are at least m_ν-ply ramified above q disjoint simply connected regions Δ_ν of a finite base surface.

Proof. For a regular exhaustion $S \to \hat{S}$ we have

$$-m_\nu \sum e(D_i) \leq n(\Delta_\nu) \leq M(\Delta_\nu)$$

for all islands D_i above Δ_ν. It follows that

$$-\sum e(D_i) \leq \frac{1}{m_\nu} M(\Delta_\nu).$$

From this and from (60) we obtain

$$(e_0+q)M \leq \sum \frac{1}{m_\nu} M(\Delta_\nu)+e^+(S)+kL.$$

On the other hand,

$$M(\Delta_\nu) \leq M+\bar{k}L,$$

and therefore

$$\sum \left(1-\frac{1}{m_\nu}\right)M \leq -e_0M+e^+(S)+\bar{\bar{k}}L.$$

This proves (67).

When there are no islands above Δ_ν, the above assertion still holds if we set $m_\nu=\infty$. If every interior point of $\hat{S}$ above a point a_ν is a branch point of multiplicity $\geq m_\nu$, then $\hat{S}$ is said to be *ramified at least m_ν-ply above a_ν*. The theorem continues to hold if we replace Δ_ν by a point a_ν because for a simply connected region Δ_ν containing a_ν every island D_i above Δ_ν has at least m_ν sheets.

§4. MEROMORPHIC FUNCTIONS

Thus far in this chapter we have considered covering surfaces from a purely metric-topological viewpoint. We now turn to covering surfaces considered as (multisheeted) Riemannian images of Riemann surfaces under analytic mappings into other Riemann surfaces. In the present section we restrict our attention to the classical case of meromorphic functions in the finite or infinite disk $|z|<\rho\leq\infty$ with values in the Riemann sphere.

In contrast with the counting function and the proximity function, the mean sheet numbers of islands and peninsulas do not involve integration. For this reason the fundamental inequality applied to meromorphic functions is called the *nonintegrated form* of the second main theorem, and Theorem I.9B (and I.18A) is referred to as the *integrated form*. In accordance with the plan of this book we shall derive the nonintegrated form *ab ovo*, for independent readability of the present chapter.

We start with a study of regular exhaustibility, then apply the fundamental inequality and give some related theorems on the existence of 1-sheeted islands and on pairs of meromorphic functions. The section closes with a second main theorem localized to a transcendental singularity of the inverse function and a similarly localized Picard theorem.

12. Regular exhaustibility

12A. Let f be meromorphic in the region R: $|z|<\rho\le\infty$, with values on the Riemann sphere S_0 of diameter 1 and tangent to the ζ-plane endowed with the spherical distance as the metric. Consider the Riemannian image $\hat{S}$ of R under f above S_0. Clearly the Riemannian image S_r of the disk $|z|<r\ (<\rho)$ is a finite covering surface of S_0 which exhausts $\hat{S}$ as $r\to\rho$. In terms of the area $A(r)$ of S_r the mean sheet number $M(r)$ of S_r is

$$M(r) = \frac{1}{\pi}A(r), \quad \text{where} \quad A(r) = \iint_{|z|<r} \frac{|f'(z)|^2}{(1+|f(z)|^2)^2}\, t\, dt\, d\theta$$

with $z=te^{i\theta}$. The length of the boundary of S_r is

$$L(r) = \int_{|z|=r} \frac{|f'(z)|}{1+|f(z)|^2}\,|dz|.$$

By the Schwarz inequality we have

$$L(r)^2 \le \int_{|z|=r} |dz| \cdot \int_{|z|=r} \frac{|f'(z)|^2}{(1+|f(z)|^2)^2}\,|dz| = 2\pi r\cdot\frac{dA(r)}{dr},$$

that is,

(68) $$\frac{dr}{r} \le 2\pi\,\frac{dA(r)}{L(r)^2}.$$

Lemma. *The Riemannian image $\hat{S}$ of $|z|<\rho$ is always regularly exhaustible for $\rho=\infty$. In the case $\rho<\infty$ the condition*

(69) $$\limsup_{r\to\rho}\,(\rho-r)A(r) = \infty$$

assures this property.

12B. For the proof consider first the parabolic case $\rho=\infty$.
On integrating (68) from $r_0>0$ to r we obtain

$$\log\frac{r}{r_0} \le 2\pi\int_{Ar_0}^{r}\frac{dA(r)}{L(r)^2},$$

which for $r\to\infty$ yields

(70) $$\int_{r_0}^{\infty}\frac{dA(r)}{L(r)^2} = \infty.$$

Suppose there were positive constants r_0 and k such that for $r_0\le r<\infty$

$$M(r) \le \frac{k}{\pi}L(r).$$

Then we would have

$$\int_{r_0}^{r}\frac{dA(r)}{L(r)^2} \le k^2\int_{r_0}^{r}\frac{dA(r)}{A(r)^2} \le \frac{k^2}{A(r_0)} < \infty,$$

in conflict with (70). We conclude that $\hat{S}$ is regularly exhaustible.

12C. Continuing with the parabolic case we insert here a related estimate. Let $\Phi(t)$ be a positive continuous function defined for $t>0$ such that

$$\int_{A(r_0)}^{\infty} \frac{dt}{\Phi(t)} < \infty.$$

Denote by E_r the set of values r with $r_0 \leq r < \infty$ which satisfy the inequality

$$L(r) \geq \sqrt{\Phi(A(r))}.$$

Then by (68)

$$\int_{E_r} d \log r \leq 2\pi \int_{E_r} \frac{dA(r)}{\Phi(A(r))} \leq 2\pi \int_{A(r_0)}^{\infty} \frac{dt}{\Phi(t)} < \infty.$$

It follows that

$$L(r) < \sqrt{\Phi(A(r))}$$

for every r which does not belong to the set E_r of finite logarithmic measure.

In particular, if we take

$$\Phi(t) = t^{1+2\varepsilon}$$

with $\varepsilon > 0$, then

$$L(r) < A(r)^{\frac{1}{2}+\varepsilon}$$

for all $r \notin E_r$. This estimate will be used in 13A.

12D. We proceed to the hyperbolic case $\rho < \infty$. Suppose that $M(r) \leq k\pi^{-1}L(r)$ for $r_0 < r < \rho$ and some constant k. Then

$$\log \frac{\rho}{r} \leq 2\pi k^2 \int_r^{\rho} \frac{dA(r)}{A(r)^2} \leq \frac{2\pi k^2}{A(r)}.$$

Consequently,

$$A(r) \leq \frac{2\pi k^2}{\log (\rho/r)} < \frac{2\pi k^2 \rho}{\rho - r},$$

that is,

$$A(r) = O\left(\frac{1}{\rho - r}\right).$$

We infer that $\mathfrak{S}$ is regularly exhaustible if (69) is satisfied. The lemma is herewith proved.

13. Application of the fundamental inequality

13A. Let Δ be a simply connected region bounded by a regular curve on the Riemann sphere S_0. Denote by $n(r, \Delta)$ the sum of the sheet numbers of the islands of S_r above Δ and by $\mu(r, \Delta)$ the sum of the mean sheet numbers of the peninsulas of S_r above Δ. Then by (48)

$$n(r, \Delta)+\mu(r, \Delta) = M(r)+O(L(r)). \tag{71}$$

Take $q \geq 3$ disjoint simply connected regions $\Delta_1, \cdots, \Delta_q$ on S_0. Let $b(r, \Delta_\nu)$ be the sum of the orders of the branch points of the islands of S_r. By (62) and 12C we have Ahlfors' [11] nonintegrated form of the second main theorem:

Theorem. *The inequality*

$$(q-2)M(r) < \sum n(r, \Delta_\nu)-\sum b(r, \Delta_\nu)+O(L(r)) \tag{72}$$

holds for all meromorphic functions in $|z|<\rho\leq\infty$. *Regular exhaustibility is here governed by Lemma* 12A. *If* $\rho=\infty$, *then*

$$L(r) = O(M(r)^{\frac{1}{2}+\varepsilon})$$

with an exceptional set E_r *of finite logarithmic measure.*

We introduce the defect and the ramification index of Δ:

$$\gamma(\Delta) = 1-\limsup_{r\to\infty} \frac{n(r, \Delta)}{M(r)} \quad \text{and} \quad \varepsilon(\Delta) = \liminf_{r\to\infty} \frac{b(r, \Delta)}{M(r)}.$$

By virtue of (72)

$$\sum \gamma(\Delta_\nu)+\sum \varepsilon(\Delta_\nu) \leq 2 \tag{73}$$

for regularly exhaustible $\hat{S}$, and (67) gives

$$\sum \left(1-\frac{1}{m_\nu}\right) \leq 2. \tag{74}$$

13B. In the case where Δ_ν is a point a_ν, and $\rho = \infty$, (72) yields

$$(q-2)M(r) < \sum n(r, a_\nu)-\sum b(r, a_\nu)+O(M(r)^{\frac{1}{2}+\varepsilon}) \tag{75}$$

with an exceptional set of values r of finite logarithmic measure. The inequality corresponds to the following integrated form of the second main theorem due to Nevanlinna [22]:

$$(q-2)C(r) < \sum A(r, a_\nu)-\sum A_1(r, a_\nu)+O(\log rC(r)). \tag{75$'$}$$

However, this form can be obtained from the nonintegrated form (72) by using a technique due to Dinghas [7], although the resulting remainder term is somewhat less sharp:

$$(q-2)C(r) < \sum A(r, a_\nu)-\sum A_1(r, a_\nu)+O(\sqrt{C(r)}\log C(r)). \tag{75$''$}$$

The exceptional set E_r of values r is again of finite logarithmic measure.

In 18 we shall generalize this result of Dinghas to the class R_p of Riemann surfaces.

13C. We return to (74) and consider, in particular, the existence of 1-sheeted islands. The following extension of Bloch's theorem [6] is due to Ahlfors [11]:

Given five disjoint simply connected regions Δ_ν of the extended plane S_0, every simply connected parabolic $\hat{S}$ above S_0 has at least one 1-sheeted island above some Δ_ν.

In fact, the regular exhaustibility of $\hat{S}$ implies that it can be completely ramified above at most four regions Δ_ν.

13D. For Riemannian images under *entire functions* there is an analogous result:

Given three disjoint simply connected regions Δ_ν of the finite plane S': $|\zeta| < \infty$, there exists at least one 1-sheeted island above some Δ_ν on every simply connected parabolic $\hat{S}$ above S'.

For the proof we observe that for a region Δ containing $\zeta = \infty$ the corresponding m in (74) is ∞.

The following result is an immediate consequence:

A covering surface generated by an entire function contains an arbitrarily large 1-sheeted circular disk.

13E. As a further consequence of Ahlfors' main theorem 6A we append here a theorem on pairs of meromorphic functions. Let

$$\Phi(z, \zeta) = 0 \tag{76}$$

be an irreducible algebraic equation, and consider the Riemann surface R_0, spread above the Riemann z-sphere, of the algebraic function defined by (76). The following classical theorem is due to Picard:

If the Euler characteristic e_0 of R_0 is positive, it is not possible to find two meromorphic functions f and g such that $\Phi(f(t), g(t)) = 0$ identically in the finite t-plane.

Suppose in fact there are two functions $z = f(t)$, $\zeta = g(t)$ meromorphic in $|t| < \infty$ which satisfy (76) identically. Denote by $\hat{R}$ the covering surface of the Riemann z-sphere generated by $z = f(t)$ as the Riemannian image of $|t| < \infty$. Clearly $\hat{R}$ is a simply connected covering surface of the base surface R_0 spread above the Riemann z-sphere. Using the spherical metric we apply Ahlfors' main theorem to an exhausting surface $R_r \to \hat{R}$:

$$e^+ \geq M(r)e_0 - kL(r).$$

Since $e^+ = 0$, we have

$$M(r) \leq \frac{k}{e_0} L(r),$$

which contradicts the fact that $\hat{R}$ is regularly exhaustible.

14. Role of the inverse function

14A. Before continuing in §5 the main train of thought of §§3–4 we digress in the remainder of the present section to the intriguing problem of relating our considerations to the inverse of a meromorphic function.

Let $\zeta=f(z)$ be nonrational and meromorphic in $|z|<\infty$. It is well known that the inverse $z=\varphi(\zeta)$ does not necessarily have a transcendental singularity. Suppose now there exists such a singularity σ above $\zeta=a$. We may assume that a is finite. We consider the covering surface $\hat{S}$ of the function φ above the extended ζ-plane. The d-neighborhood S_d of the accessible boundary point σ of $\hat{S}$ is a covering surface of the base disk D: $|\zeta-a|<d$. The function φ maps S_d bijectively onto a region G in the z-plane. The boundary Γ of G consists of at most countably many analytic curves and the point $z=\infty$. We note that f is analytic on the closure $\bar{G}$ except at $z=\infty$ and that $|f(z)-a|=d$ on Γ except at $z=\infty$. Moreover, there exists a path Λ in G terminating at $z=\infty$ along which f tends to a (Iversen [1]).

Choose a positive number r_0 such that the circle $|z|=r_0$ contains the finite end point of Λ and at least one boundary point of G in its interior. For $r_0\le r<\infty$ the intersection Θ_r of the circle K_r: $|z|=r$ and the region G consists of finitely many circular arcs. We denote by G_r the intersection of the interior of K_r and G, and by $A(r)$ the Euclidean area of the Riemannian image $S(r)$ of G_r under f. We shall show:

Lemma. *The area* $A(r)$ *is unbounded,*

$$\lim_{r\to\infty} A(r) = \infty. \tag{77}$$

Proof. If for all sufficiently small d the relative boundary Γ of G consists of only bounded closed curves, then the assertion is true because the inverse φ has no transcendental singularities other than σ above the disk D: $|\zeta-a|<d$. For the proof of (77) it suffices to consider the case where the relative boundary Γ *contains at least one boundary curve starting from* $z=\infty$ *and terminating at* $z=\infty$.

We denote by $t(r)$ the last point of intersection of Λ with the circle K_r starting from the initial point t_0 of Λ, and by Λ_r the part of Λ from the point $t(r)$ to $z=\infty$. There exists a positive number $r_1>r_0$ such that

(a) the image of the curve Λ_{r_1} under f is contained in the disk $D_{1/2}$: $|\zeta-a|<d/2$,

(b) if $r_1\le r<\infty$, the set Θ_r contains a cross-cut $\gamma(r)$ of G which contains the point $t(r)$ in its interior, and whose end points $t'(r)$ and $t''(r)$ are on Γ.

We note that the image of the cross-cut $\gamma(r)$ under f is a curve starting from a point on the circle C: $|\zeta-a|=d$, passing through a point in $D_{1/2}$

and terminating at a point on C. Consequently the length of the image of $\gamma(r)$ is at least d.

Denote by $r\theta(r)$ and $L(r)$ the total lengths of Θ_r and the image of Θ_r under f, respectively. Then $L(r)=\int_{\Theta_r}|f'(z)|r\,d\theta$, where $z=re^{i\theta}$. By the Schwarz inequality we obtain

$$L(r)^2 \leq \int_{\Theta_r} r\,d\theta \cdot \int_{\Theta_r} |f'(z)|^2 r\,d\theta,$$

whence

$$\frac{L(r)^2}{r\theta(r)} \leq \int_{\Theta_r} |f'(z)|^2 r\,d\theta \tag{78}$$

and

$$\int_{r_1}^{r} \frac{L(r)^2}{r\theta(r)}\,dr \leq \int_{r_1}^{r}\int_{\Theta_r} |f'(z)|^2 r\,dr\,d\theta = A(r)-A(r_1).$$

Since for $r_1 \leq r < \infty$, $r\theta(r) \leq 2\pi r$, and $d \leq L(r)$, we have

$$\int_{r_1}^{r} \frac{L(r)^2}{r\theta(r)}\,dr \geq \frac{d^2}{2\pi}\int_{r_1}^{r} d\log r = \frac{d^2}{2\pi}\log\frac{r}{r_1}.$$

This proves the lemma.

15. Localized second main theorem

15A. If the circle K_r is contained in G for $r_0 \leq r < \infty$, then the set G_r consists of a single component $G^1(r)$ bounded by analytic contours Γ_1^λ, $\lambda=1,\cdots,k$, contained in the relative boundary Γ of G, and the cycle K_r in G. It is clear that f maps each Γ_1^λ onto a curve on C while K_r is mapped onto a curve in D. Except for the case where K_r is a cycle of G the open set G_r consists of finitely many components

$$G^1(r),\cdots,G^m(r),$$

where $m=m(r)\geq 1$.

The boundary of each component $G^i(r)$, $i=1,\cdots,m$, consists of at least one analytic curve Γ_i^λ, $\lambda=1,\cdots,l$, contained in Γ and at least one cross-cut γ_i^μ, $\mu=1,\cdots,h$. The function f maps each Γ_i^λ onto a curve on C and each γ_i^μ onto a curve in D except for the two end points on C.

In both cases the Riemannian image $S^i(r)$ of $G^i(r)$ under f is a finite covering surface of the base surface D. On taking $q\geq 2$ disjoint circular disks $D_1,\cdots,D_q$ in the interior of D, applying (63) to each $S^i(r)$, and setting $e_0=-1$ we obtain for $i=1,\cdots,m$

$$\sum[M^i-n^i(D_j)] \leq M^i-\sum b^i(D_j)+e^+(S^i)+kL^i, \tag{79}$$

where k is a constant depending only on the disks D_j. Summation of the m inequalities gives

$$(80)\qquad \sum_{i=1}^{m}\sum_{j=1}^{q}[M^i-n^i(D_j)] \le \sum_{i=1}^{m} M^i-\sum_{i=1}^{m}\sum_{j=1}^{q} b^i(D_j)+\sum_{i=1}^{m} e^+(S^i)+k\sum_{i=1}^{m} L^i.$$

15B. Denote by ν^i the number of holes in $G^i(r)$, i.e., the number of contours of $G^i(r)$ which contain no points of K_r. Clearly

$$e^+(G^i(r)) = e^+(S^i) \le e(S^i)+1 = \nu^i.$$

Since ν^i is at most the mean sheet number $M^i(C)$ of the compact bordered surface $\bar{S}^i$ above C, we have by Theorem 3A

$$(81)\qquad \nu^i \le M^i(C) \le M^i+k'L^i.$$

We introduce the following notations:

$$\sum_{i=1}^{m(r)} n^i(D_j) = n(r, D_j, G), \qquad \sum_{i=1}^{m(r)} b^i(D_j) = b(r, D_j, G),$$

$$\sum_{i=1}^{m(r)} M^i = M(r, G), \qquad \sum_{i=1}^{m(r)} L^i = L(r, G),$$

$$\sum_{i=1}^{m(r)} \nu^i = \nu(r, G),$$

where $\nu(r, G)$ is the number of holes in G encircled by K_r. In terms of these notations we obtain from (80) the following inequality (Noshiro [2]):

Theorem. *The estimate*

$$(82)\qquad \sum (M(r, G)-n(r, D_j, G)) \le M(r, G)-\sum b(r, D_j, G)+\nu(r, G)+kL(r, G)$$

holds for a neighborhood D of a transcendental singularity of the inverse function of a function meromorphic in $|z|<\infty$.

16. Localized Picard theorem

16A. Clearly by (81)

$$(83)\qquad \nu(r, G) \le M(r, G)+k'L(r, G).$$

We introduce the quantities

$$(84)\qquad \bar{\eta} = \limsup_{r\to\infty} \frac{\nu(r, G)}{M(r, G)}, \quad \eta = \liminf_{r\to\infty} \frac{\nu(r, G)}{M(r, G)},$$

Since $\lim_{r\to\infty} M(r, G)=\infty$, $\bar{\eta}$ and η are the orders of growth of the number of holes of G.

We define the defect $\gamma(D_j, G)$ and the ramification index $\varepsilon(D_j, G)$ of the disk D_j relative to G by

$$\gamma(D_j, G) = 1 - \limsup_{r\to\infty} \frac{n(r, D_j, G)}{M(r, G)},$$

$$\varepsilon(D_j, G) = \liminf_{r\to\infty} \frac{b(r, D_j, G)}{M(r, G)}.$$

16B. We shall prove that the exhaustion $S(r) \to S_d$ is regular. We start from (78),

$$\frac{dr}{r\theta(r)} \le \frac{dA(r)}{L(r)^2},$$

and denote by E_r the set of $r \ge r_0$ such that $L(r) \ge A(r)^{\frac{1}{2}+\varepsilon}$, $\varepsilon > 0$. Then

$$\frac{1}{2\pi}\int_{E_r} d\log r \le \int_{E_r} \frac{dr}{r\theta(r)} \le \int_{E_r} \frac{dA(r)}{A(r)^{1+2\varepsilon}} \le \int_{A(r_0)}^{\infty} \frac{dt}{t^{1+2\varepsilon}} < \infty.$$

Hence $L(r) < A(r)^{\frac{1}{2}+\varepsilon}$ for $r \notin E_r$, and we conclude that

$$\liminf_{r\to\infty} \frac{L(r, G)}{M(r, G)} = 0.$$

In particular, this regular exhaustibility together with (83) implies

(85) $$0 \le \eta \le 1.$$

16C. From (82) and (83) we obtain

(86) $$\sum \gamma(D_j, G) + \sum \varepsilon(D_j, G) \le \min(2, 1+\bar{\eta}).$$

As a special case we have the following localized Picard theorem due to Noshiro [2], Kunugui [1], [3], and Tumura [4], [5].

Let f be nonrational and meromorphic in $|z| < \infty$. Suppose that the inverse φ of f has a transcendental singularity σ above $\zeta = a$ and denote by $\hat{S}$ the Riemann surface of φ. A d-neighborhood S_d of the accessible boundary point σ is a covering surface of the base surface D: $|\zeta - a| < d$. Let G be the image of S_d under φ.

Theorem. *The function f takes on every value in D infinitely often in G except for at most two values. In particular, if G is simply connected (or of finite multiple connectivity), then f assumes every value of D infinitely often in G with one possible exception.*

16D. Under the assumption of this theorem, if f does not take the value a in the region $G = G_d$ for a sufficiently small d, then σ is said to be a *direct transcendental singularity* of φ.

Corollary. *The inverse φ has at most countably many direct transcendental singularities.*

Proof. Let $\hat{S}$ be the Riemann covering surface generated by f above the extended ζ-plane. By the above theorem the part of $\hat{S}$ above a disk D: $|\zeta - a| < d$ consists of at most countably many components $\hat{S}_\nu$ which cover every point of D at least once except for at most two. Let D_n: $|\zeta - a_n| < d_n$, $n = 1, 2, \cdots$, be disks with rational a_n, d_n. Denote by E_n the set of points of D_n which are not covered by at least one component of the part of $\hat{S}$ above D_n. Obviously $E_\infty = \bigcup_1^\infty E_n$ is at most countable. Let σ be a direct transcendental singularity of φ above a. Then a is not covered by at least one component of the part of $\hat{S}$ above some D_n, so that $a \in E_n$. Hence $a \in E_\infty$ and we infer that φ has at most countably many direct transcendental singularities.

For an extension of this corollary we refer to Tsuji [19].

§5. MAPPINGS OF ARBITRARY RIEMANN SURFACES

From the special case of meromorphic functions we proceed to given analytic mappings f of arbitrary Riemann surfaces R into closed Riemann surfaces S_0. For preparation we derive a condition for the Riemannian image of R above S_0 to be regularly exhaustible. The chapter culminates in the nonintegrated form 18C of the second main theorem for arbitrary Riemann surfaces.

We then show that for R_p-surfaces (I.12 and IV) an integrated form can be derived from this nonintegrated form. This general case is compared with the concrete case of algebroids. The chapter closes with a sharpness proof of the generalized defect relation in the nonintegrated form.

17. Conformal metrics

17A. We consider a given analytic mapping f of an arbitrary Riemann surface R into a closed Riemann surface S_0.

On R let

$$d\rho = \mu(z)|dz| \tag{87}$$

be a conformally invariant metric. (While in §4 we used the symbol r in its traditional meaning, and will do so for algebroids in 20, we interchange here r and ρ to reserve the standard notation $d\rho$ for the metric.) For a given $z_0 \in R$ the distance $\rho(z, z_0) \geq 0$ between $z \in R$ and z_0 is defined as $\inf \int_\alpha d\rho$ over all rectifiable arcs from z to z_0. Set $\rho_\beta = \sup_{z \in R} \rho(z, z_0)$.

For every open Riemann surface there exists a conformal metric such that $\rho_\beta = \infty$ *and the set*

(88) $$\beta_r = \{z \mid \rho(z, z_0) = r\}$$

is compact for every $r > 0$.

In fact, a conformal mapping of the universal covering surface R^∞ of R onto the disk $|w| < 1$ gives the hyperbolic metric $d\rho = |dw(z)|/(1-|w(z)|^2)$ with the desired properties. The degenerate case where R^∞ is conformally equivalent to the plane $|w| < \infty$ only occurs if R is simply or doubly connected and parabolic. Then we can use the capacity metric

(89) $$d\rho = \frac{1}{2\pi} |\text{grad } p_\beta| \, |dz|,$$

where p_β is the capacity function with compact level lines (I.12 and IV).

In case R is the interior of a compact bordered Riemann surface, a metric with $\rho_\beta < \infty$ is also available, e.g., by (89), and is perhaps a more natural choice. For this reason we shall discuss both cases $\rho_\beta < \infty$ and $\rho_\beta = \infty$.

17B. On the closed range surface S_0 we choose two points ζ_0, ζ_1 and disjoint parametric disks D_0, D_1 about ζ_0, ζ_1. Let t_0 be a harmonic function on $S_0 - \zeta_0 - \zeta_1$ with singularities $-2 \log |\zeta - \zeta_0|$ and $2 \log |\zeta - \zeta_1|$ in D_0 and D_1, respectively. We normalize the additive constant by the condition $t_0(\zeta) + 2 \log |\zeta - \zeta_0| \to 0$ as $\zeta \to \zeta_0$. As in I.1 we use the function

(90) $$s_0 = \log (1 + e^{t_0})$$

which has a positive logarithmic pole at ζ_0 and is nonnegative on S_0.

On S_0 we consider the metric

(91) $$d\sigma = \lambda(\zeta) |d\zeta|,$$

where

(92) $$\lambda^2 = \Delta s_0 = \frac{e^{t_0} |\text{grad } t_0|^2}{(1 + e^{t_0})^2}.$$

It is easily seen that λ has no singularities (cf. II.8C) and that the total area of S_0 is

(93) $$\int_{S_0} \lambda^2 dS = 4\pi,$$

where dS is the Euclidean area element in a parametric disk of S_0 (I.1C).

18. Main theorem for arbitrary Riemann surfaces

18A. Given an analytic mapping f of R into S_0 set

$$f_\rho = \frac{df/dz}{d\rho/|dz|} = f'\mu^{-1}.$$

The length of the image of β_r is

$$L(r) = \int_{\beta_r} \lambda |f_\rho| \, d\rho, \tag{94}$$

and the mean sheet number $M(r)$ of the Riemannian image of $R_r = \{z \mid \rho(z, z_0) < r\}$ has the derivative

$$\frac{dM(r)}{dr} = \frac{1}{4\pi} \int_{\beta_r} \lambda^2 |f_\rho|^2 \, d\rho. \tag{95}$$

We denote the length of β_r by $l(r) = \int_{\beta_r} d\rho$.

Lemma. *For* $0 < r < \rho_\beta$

$$\frac{dr}{l(r)} \leq 4\pi \frac{dM(r)}{L(r)^2}. \tag{96}$$

This is a direct consequence of the Schwarz inequality

$$L(r)^2 \leq \int_{\beta_r} d\rho \cdot \int_{\beta_r} \lambda^2 |f_\rho|^2 d\rho.$$

18B. We can now state a general condition for regular exhaustibility (Sario [7]):

Lemma. *The Riemannian image of an arbitrary Riemann surface under an analytic mapping into a closed Riemann surface has the property*

$$\liminf_{\rho \to \rho_\beta} \frac{L(\rho)}{M(\rho)} = 0, \tag{97}$$

if $M(\rho)$ *grows so rapidly that*

$$\limsup_{\rho \to \rho_\beta} M(\rho) \int_\rho^{\rho_\beta} \frac{dr}{l(r)} = \infty. \tag{98}$$

Proof. Suppose the conclusion were not true: $\liminf (L/M) > 0$ and there existed a ρ_0 with $0 < \rho_0 < \rho_\beta$ such that $L(r) > qM(r)$ for $\rho_0 < r < \rho_\beta$. It would follow that

$$\int_\rho^{\rho_\beta} \frac{dr}{l(r)} \leq 4\pi \int_\rho^{\rho_\beta} \frac{dM(r)}{L(r)^2} < \frac{4\pi}{q^2} \int_\rho^{\rho_\beta} \frac{dM(r)}{M(r)^2} \leq \frac{4\pi}{q^2} \frac{1}{M(\rho)},$$

and therefore

$$M(\rho)\int_{\rho}^{\rho_\beta}\frac{dr}{l(r)} \le \frac{4\pi}{q^2},$$

in violation of (98).

18C. Let $e(\rho)$ be the Euler characteristic of R_ρ. We apply the generalization (51) of Ahlfors' fundamental inequality to the Riemannian image of R under f and collect our results in the following main statement (Sario [7], Noshiro):

Theorem. *For an analytic mapping of an arbitrary Riemann surface into a closed Riemann surface,*

$$(99)\qquad (e_0+q)M(\rho) < \sum n(\rho,\Delta_\nu)-\sum b(\rho,\Delta_\nu)+e^+(\rho)+O(L(\rho)).$$

Here the condition lim inf $(L/M)=0$ *of regular exhaustibility is satisfied by mappings with property* (98).

In the definitions 11A of the defect $\gamma(\Delta_\nu)$, the ramification index $\varepsilon(\Delta_\nu)$ and the Euler index ξ, the directed limits can now be replaced by ordinary limits for $\rho \to \rho_\beta$ as the Riemannian image S_ρ of R_ρ exhausts that of R. For regular exhaustions the defect and ramification relation

$$\sum \gamma(\Delta_\nu)+\sum \varepsilon(\Delta_\nu) \le \xi - e_0,$$

and the relation

$$\sum\left(1-\frac{1}{m_\nu}\right) \le \xi - e_0$$

on m_ν-ply ramified coverings remain valid in the present setup of analytic mappings of arbitrary Riemann surfaces into closed Riemann surfaces.

For an illustration of (99) and its consequences we refer to the concrete case of Gaussian mappings (App. II.7–8).

18D. We consider, in particular, R_p-surfaces characterized by the existence of a capacity function p_β with compact level lines (I.12 and IV). Recall that, e.g., all parabolic and all compact bordered surfaces have this property. It is possible to choose the parametric disk at the pole of p_β such that sup $p_\beta>0$. We change our notations slightly by setting for $0\le\rho<\rho_\beta=\sup p_\beta$,

$$(100)\qquad \beta_\rho = \{z \mid p_\beta(z) = \rho\}, \qquad R_\rho = \{z \mid p_\beta(z) < \rho\},$$

and choose ρ so that grad $p_\beta \ne 0$ on β_ρ. Since $l(r)=\int_{\beta_r} d\rho = 1$, the negligibility of the remainder in (99) takes the following form:

The condition lim inf $(L/M)=0$ *is satisfied by all analytic mappings of parabolic surfaces and by those analytic mappings of hyperbolic R_p-surfaces for which*

$$(101)\qquad \limsup_{\rho\to\rho_\beta} M(\rho)(\rho_\beta-\rho) = \infty.$$

19. Integrated form

19A. We shall show that for R_p-surfaces an integrated form, analogous to I.(47) and I.(102), can be derived from the nonintegrated form (99) of the main theorem.

We introduce the integrated quantities

$$C(\rho) = \int_0^\rho M(r)\,dr, \qquad A(\rho, \Delta) = \int_0^\rho n(r, \Delta)\,dr,$$

$$A_1(\rho, \Delta) = \int_0^\rho b(r, \Delta)\,dr, \qquad E(\rho) = \int_0^\rho e^+(r)\,dr,$$

and

$$L^*(\rho) = \int_0^\rho L(r)\,dr.$$

We first obtain

$$(e_0+q)C(\rho) \leq \sum A(\rho, \Delta_\nu) - \sum A_1(\rho, \Delta_\nu) + E(\rho) + O(L^*(\rho)). \tag{102}$$

The problem is to find a majorant for $L^*(\rho)$ in terms of the characteristic function $C(\rho)$. We note that $l(r) \equiv 1$ and obtain by (96)

$$L^*(\rho)^2 \leq \rho \int_0^\rho L(r)^2 dr < 4\pi\rho \left[\frac{dC(r)}{dr}\right]_{r=\rho}$$

for $0 \leq \rho < \rho_\beta$. On a parabolic surface R_p every nonconstant mapping f has an unbounded characteristic $C(\rho)$. On a hyperbolic R_p we only consider mappings with this property.

19B. To obtain concrete results we shall estimate L^* making use of the following special function, although the reasoning can be carried out analogously with more general convex functions. For an integer $i \geq 1$ and a function $\varphi(\rho)$ denote by $\log^{(i)}\varphi(\rho)$ the ith iterate of the logarithm and set

$$P_j(\varphi(\rho)) = \begin{cases} \prod_1^j \log^{(i)}\varphi(\rho) & \text{for } j \geq 1, \\ 1 & \text{for } j = 0. \end{cases}$$

We start with the *parabolic* case and claim that for given integers n, $m \geq 1$

$$L^*(\rho) < \sqrt{\frac{C(\rho)P_{m-1}(C(\rho))}{P_{n-1}(\rho)}}\,\log^{(m)}C(\rho) \tag{103}$$

except in a set $\Delta_{nm} \subset (0, \infty)$ such that $\int_{\Delta_{nm}} d\log^{(n)}\rho < \infty$.

For the proof choose any $\rho_0 > 0$ and let

$$\Delta_{nm} = \{\rho \mid \rho > \rho_0, \text{(103) false}\}.$$

Then

$$\int_{\Delta_{nm}} d\log^{(n)}\rho \le 4\pi\int_{\rho_0}^{\infty}\frac{dC(\rho)}{C(\rho)P_{m-1}(C(\rho))(\log^{(m)}C(\rho))^2}$$
$$= \frac{4\pi}{\log^{(m)}C(\rho_0)} < \infty.$$

19C. Next suppose R_p is *hyperbolic.* We claim that

$$L^*(\rho) < \sqrt{\frac{C(\rho)P_{m-1}(C(\rho))}{(\rho_\beta-\rho)P_{n-1}(1/(\rho_\beta-\rho))}}\log^{(m)}C(\rho) \tag{104}$$

except in a set $\Delta_{nm}\subset(0,\rho_\beta)$ such that $\int_{\Delta_{nm}} d\log^{(n)}(1/(\rho_\beta-\rho))<\infty$.

For $0<\rho_0<\rho_\beta$ let

$$\Delta_{nm} = \{\rho \mid \rho_0<\rho<\rho_\beta,\ (104)\ \text{false}\}.$$

Then

$$\int_{\Delta_{nm}} d\log^{(n)}\frac{1}{\rho_\beta-\rho} \le 4\pi\rho_\beta\int_{\rho_0}^{\rho_\beta}\frac{dC(\rho)}{C(\rho)P_{m-1}(C(\rho))(\log^{(m)}C(\rho))^2} < \infty.$$

19D. We have shown that (99) implies the following integrated form of the main theorem, which we state in a somewhat simpler but less accurate form than (103), (104) would permit:

Theorem. *Under an analytic mapping of an R_p-surface into a closed Riemann surface S_0*

$$(q+e_0)C(\rho) < \sum A(\rho,\Delta_\nu)-\sum A_1(\rho,\Delta_\nu)+E(\rho)+F(\rho). \tag{105}$$

For a parabolic R_p-surface the remainder $F(\rho)$ is

$$F(\rho) = O(\sqrt{C(\rho)}\log C(\rho)) \tag{106}$$

except in a set Δ so small that $\int_\Delta d\log\rho<\infty$. For a hyperbolic R_p-surface

$$F(\rho) = O\left(\sqrt{\frac{C(\rho)}{\rho_\beta-\rho}}\log C(\rho)\right) \tag{107}$$

except in a set Δ such that $\int_\Delta d\log(\rho_\beta-\rho)^{-1}<\infty$.

Remark. The theorem is due to Noshiro-Sario [1]. In the case of the plane R_p and the extended plane S_0, (106) is the interesting result of Dinghas [7] (cf. (75)″).

Clearly our theorem can also be stated in the weaker form

$$F(\rho) = O(C(\rho)^{\frac{1}{2}+\varepsilon})$$

if R_p is parabolic, and

$$F(\rho) = O((\rho_\beta-\rho)^{-\frac{1}{2}}C(\rho)^{\frac{1}{2}+\varepsilon})$$

if R_p is hyperbolic.

19E. We introduce the "integrated forms" of the defect

$$\delta(\Delta_\nu) = 1 - \limsup_{\rho \to \rho_B} \frac{A(\rho, \Delta_\nu)}{C(\rho)},$$

the ramification index

$$\vartheta(\Delta_\nu) = \liminf_{\rho \to \rho_B} \frac{A_1(\rho, \Delta_\nu)}{C(\rho)},$$

and the Euler index

$$\kappa = \limsup_{\rho \to \rho_B} \frac{E(\rho)}{C(\rho)}.$$

The defect and ramification relation reads:

Corollary. *The bound*

(108) $$\sum \delta(\Delta_\nu) + \sum \vartheta(\Delta_\nu) \le \kappa - e_0$$

holds for every f on a parabolic R_p and for those f on a hyperbolic R_p for which

(109) $$\lim_{\rho \to \rho_B} \frac{\log C(\rho)}{\sqrt{(\rho_B - \rho)C(\rho)}} = 0.$$

Less sharply, it suffices in the latter case that

(109)′ $$\lim_{\rho \to \rho_B} (\rho_B - \rho)C(\rho)^{1-\varepsilon} = \infty$$

for some $\varepsilon > 0$.

Remark. As in 10D we can replace the regions Δ_ν by points a_ν in 19D and 19E. In the special case of meromorphic functions in the finite or infinite disk $|z|(=e^\rho) < e^{\rho_B} \le \infty$, (108) reduces to the classical defect and ramification relation $\sum \delta(a_\nu) + \sum \vartheta(a_\nu) \le 2$.

20. Algebroids

20A. A comparison of the general Theorem 19D with the corresponding theorem on algebroid functions offers considerable interest. All results in this no. are due to Noshiro.

Let f be a k-valued algebroid function in $|z| < \rho \le \infty$ and denote by R its Riemann surface above $|z| < \rho$. We suppose that R has no branch point above $z = 0$. We assume that the part $R(r)$ of R above $|z| < r$ consists of k sheets for $r_0 \le r < \rho$. (In the present no. it is again more natural to use r in its traditional meaning.) The Hurwitz formula gives for the Euler characteristic $e(r)$ of $R(r)$

$$e(r) = -k + c(r),$$

where $c(r)$ is the sum of the orders of the branch points of $R(r)$. Clearly $e^+(r) \le e(r) + 1 \le c(r)$.

Denote by $A(r)$ the *spherical area* of the Riemannian image of $R(r)$. Then by H. Selberg's theorem [9, p. 11]

$$(110) \qquad E(r) = \int_{r_0}^{r} \frac{e^+(r)}{r}\,dr < \frac{2k-2}{\pi}\int_{r_0}^{r} \frac{A(r)}{r}\,dr + O(1).$$

20B. Suppose that S_0 is a closed Riemann surface above the Riemann ζ-sphere, and that the Riemannian image $\hat{S}$ of R under the algebroid function f is a covering surface of S_0. By (99) we have

$$(e_0+q)M(r) < \sum n(r, \Delta_\nu) - \sum b(r, \Delta_\nu) + e^+(r) + O(L(r)).$$

On dividing by r and integrating from r_0 to r we obtain, with the notations of 19A modified in an obvious manner,

$$(e_0+q)C(r) < \sum A(r, \Delta_\nu) - \sum A_1(r, \Delta_\nu) + E(r) + O(L^*(r)).$$

Here L^* is estimated in the same fashion as in 19A. In the present situation we set for $0 < r_0 < r < \rho$

$$C(r) = \int_{r_0}^{r} \frac{M(r)}{r}\,dr = \frac{1}{\pi N_0}\int_{r_0}^{r} \frac{A(r)}{r}\,dr,$$

where N_0 and πN_0 are the sheet number and the area of S_0. We have arrived at the following counterpart of Theorem 19D (cf. Noshiro [2]):

Theorem. *For algebroid functions*

$$(111) \quad \sum (C(r) - A(r, \Delta_\nu)) + \sum A_1(r, \Delta_\nu) < [(2k-2)N_0 - e_0]C(r) + O(L^*),$$

where $L^*(r)$ *is majorized by* (106), (107), *with exceptional intervals as in* 19D.

20C. For the integrated forms of the defect and ramification index of 19E we have:

Corollary. *The relation*

$$(112) \qquad \sum \delta(\Delta_\nu) + \sum \vartheta(\Delta_\nu) \le (2k-2)N_0 - e_0$$

holds for every algebroid on a parabolic R *and for those algebroids on a hyperbolic* R *that satisfy the condition*

$$\lim_{r\to\rho} (\rho - r)C(r)^{1-\varepsilon} = 0, \qquad \varepsilon > 0.$$

Since $e_0 = -2N_0 + b$, with b the sum of the orders of the branch points of S_0, we can also write (112) in the following form:

$$(112)' \qquad \sum \delta(\Delta_\nu) + \sum \vartheta(\Delta_\nu) \le 2kN_0 - b.$$

Note that $(2k-2)N_0 \ge e_0$ or $2kN_0 \ge b$.

20D. In analogy with 11B we obtain

$$\sum\left(1-\frac{1}{m_\nu}\right)M < -e_0M+e^+ +O(L),$$

which applied to the present case gives

$$\sum\left(1-\frac{1}{m_\nu}\right)C(r) < -e_0C(r)+(2k-2)N_0C(r)+O(L^*(r)).$$

Theorem. *Under the same assumption as above*

(113) $$\sum\left(1-\frac{1}{m_\nu}\right) \le (2k-2)N_0-e_0 = 2kN_0-b.$$

21. Sharpness of nonintegrated defect relation

21A. We return to *arbitrary* Riemann surfaces and consider meromorphic functions f on them. For the nonintegrated forms of the defect γ and the Euler index ξ we have by (99)

(114) $$\sum \gamma(\Delta_\nu) \le 2+\xi$$

with $\xi=\limsup_{\rho\to\rho_\beta} e^+(\rho)/M(\rho)$. For the corresponding integrated forms δ, κ

(115) $$\sum \delta(\Delta_\nu) \le 2+\kappa$$

with $\kappa=\limsup_{\rho\to\rho_\beta} E(\rho)/C(\rho)$, where $C(\rho)=\int_0^\rho M(\rho)\,d\rho$ and $E(\rho)=\int_0^\rho e^+(\rho)\,d\rho$.

For the sake of completeness of the present chapter we shall establish the sharpness of these bounds (cf. I.10–11). We can now make use of results in 20.

21B. We again start with the simple k-valued algebroid

(116) $$\zeta = f(z) = \sqrt[k]{\frac{e^z+i}{e^z-i}}$$

to give the proof for even integers. The Riemann surface R of the function is a k-sheeted covering surface above the finite z-plane with branch points $z_j=i(\pi/2+j\pi)$, $j=0, \pm 1, \pm 2, \cdots$, of multiplicity k. Choose on R the metric $d\rho=|dz|/2\pi k|z|$ and set $p_\beta=(2\pi k)^{-1}\log|z|$. Then β_ρ of (100) is the k-sheeted circle $|z|=e^{2\pi k\rho}$ with length $l(\rho)=(2\pi k)^{-1}\int_{\beta_\rho} d\arg z=1$. Obviously R is parabolic. By Selberg's theorem (cf. (110))

$$E(\rho) < (2k-2)C(\rho)+O(1).$$

From this and from (115) it follows that $\kappa \le 2k-2$ and $\sum \delta(\Delta_\nu) \le 2k$. Since f has exactly $2k$ exceptional values, distributed equidistantly on $|\zeta| = 1$, this shows that (115) is sharp.

21C. We shall prove that (114) is also sharp. In view of

$$\liminf_{\rho \to \infty} \frac{L(\rho)}{M(\rho)} = 0$$

we have

$$\xi = \limsup_{\rho \to \infty} \frac{e^+}{n(\Delta)+\mu(\Delta)} \le \limsup_{\rho \to \infty} \frac{e^+}{n(\Delta)},$$

where we choose for Δ a small disk about $\zeta = 0$. Then $n(\Delta)$ is the number of zeros of f. Exhaust R by k-sheeted disks R_m: $|z| < 2\pi m$, $m = 1, 2, \cdots$, and indicate quantities referring to R_m by the subindex m. Then

$$\limsup_{\rho \to \infty} \frac{e^+}{n(\Delta)} = \limsup_{m \to \infty} \frac{e_m^+}{n_m(\Delta)}. \tag{117}$$

When bounded terms are disregarded, we find from the Hurwitz formula that $e_m \sim 4m(k-1)$. The zeros of f are at $e^z = -i$, $z_j = i(-\pi/2+2\pi j)$, $j = 0, \pm 1, \pm 2, \cdots$, and therefore $n_m(\Delta) \sim 2m$. It follows that

$$\xi \le 2k-2,$$

and we conclude that (114) is sharp.

22. Direct estimate of $M(\rho)$

22A. The above proof makes essential use of (48): $n(\Delta)+\mu(\Delta) = M + O(L)$. It is illuminating that M can also be estimated directly, without invoking this relation and yields exactly the same bound for ξ.

We consider a slightly more general function (Sario [7])

$$\eta(z) = f(z)^h = \left(\frac{e^z+i}{e^z-i}\right)^{h/k}, \tag{118}$$

which has $2k/h$ exceptional values, h being a factor of $2k$. Let $R_{mj} \subset R_m$ be the k-sheeted rectangle

$$|x| < 2\pi\sqrt{m^2-j^2}, \qquad 2\pi(j-1) < y < 2\pi j,$$

$j = 1, \cdots, m-1$. If the contribution of R_{mj} to M_m is denoted by M_{mj}, then clearly

$$M_m > 2 \sum_{j=1}^{m-1} M_{mj}. \tag{119}$$

Under the mapping $s=e^z$ the rectangle R_{mj} becomes a k-sheeted annulus with outer radius

$$R = \exp(2\pi\sqrt{m^2-j^2}) \tag{120}$$

and inner radius R^{-1}. The function $t=(s+i)/(s-i)$ maps the annulus onto the k-sheeted complement of two Steiner circles encircling $t=1$ and $t=-1$, respectively, symmetrically placed about the real and imaginary t-axes and intersecting the real axis at the images of $s=\pm iR$, $\pm iR^{-1}$, i.e., at distances

$$t_1 = \frac{R+1}{R-1} \tag{121}$$

and t_1^{-1} from $t=0$. The function $\zeta=\sqrt[k]{t}$ maps the k-sheeted complement of the two Steiner disks onto the 1-sheeted complement of the $2k$ images of the disks, which appear as distorted disks encircling points $\zeta=e^{i\varphi_\nu}$, $\varphi_\nu=\nu\pi/k$, $\nu=1,\cdots,2k$, and are located in the annulus

$$\sqrt[k]{t_1^{-1}} < |w| < \sqrt[k]{t_1}. \tag{122}$$

The function $\eta=w^h$ gives as the final image $\eta(R_{mj})$ of R_{mj} the h-sheeted complement of $2k/h$ distorted disks encircling points $\eta=e^{i\alpha_\nu}$, $\alpha_\nu=\nu h\pi/k$, $\nu=1,\cdots,2k/h$, and located in the annulus

$$r_1 = t_1^{-h/k} < |\eta| < t_1^{h/k} = r_1^{-1}. \tag{123}$$

By definition the mean sheet number M_{mj} of the image of R_{mj} in the stereographic η-metric is the π^{-1}-fold area of $\eta(R_{mj})$. By omitting the annulus (123) we obtain

$$\begin{aligned} M_{mj} &> \frac{h}{\pi}\left(\int_0^{2\pi}\int_0^{r_1}\frac{r\,d\varphi\,dr}{(1+r^2)^2}+\int_0^{2\pi}\int_{r_1^{-1}}^{\infty}\frac{r\,d\varphi\,dr}{(1+r^2)^2}\right)\\ &= -h\left(\frac{1}{1+r_1^2}-1-\frac{1}{1+r_1^{-2}}\right) = \frac{2hr_1^2}{1+r_1^2}. \end{aligned}$$

On setting

$$\varepsilon_{mj} = \frac{2}{\exp(2\pi\sqrt{m^2-j^2})-1} \tag{124}$$

we have $t_1=1+\varepsilon_{mj}$ and

$$M_{mj} > \frac{2h(1+\varepsilon_{mj})^{-2h/k}}{1+(1+\varepsilon_{mj})^{-2h/k}} > h(1+\varepsilon_{mj})^{-2h/k}.$$

Here

$$\varepsilon_{mj} \leq \frac{2}{\exp(2\pi\sqrt{2m-1})-1} = \varepsilon_m$$

and we find by (119) that

$$M_m > 2\sum_{j=1}^{m-1} h(1+\varepsilon_m)^{-2h/k} = 2(m-1)h(1+\varepsilon_m)^{-2h/k}.$$

For the Euler characteristic we have as before $e_m \sim 4m(k-1)$. Hence

$$\limsup_{m\to\infty} \frac{e_m}{M_m} \le \limsup_{m\to\infty} \frac{2m(k-1)}{(m-1)h(1+\varepsilon_m)^{-2h/k}} = \frac{2(k-1)}{h}.$$

In the special case $h=1$ the value is $2(k-1)$, in perfect agreement with the result in 21C.

23. Extension to arbitrary integers

23A. In conclusion we give Rodin's [2] important modification of (116) to show that (114) is sharp for all integers $\xi \ge 0$. Take k copies of the finite z-plane, each slit along the rays

(125) $$I_j = \{z \mid x < 0, y = 2\pi j\}, \qquad j = 0, \pm 1, \pm 2, \cdots.$$

For a fixed j the edges of the slits of I_j are identified on the sheets so as to obtain a branch point of multiplicity k at $z = 2\pi ji$. This forms a covering surface R of the z-plane. As in 21B we now choose the metric $d\rho = |dz|/2\pi k|z|$, $p_\beta = (2\pi k)^{-1}\log|z|$, β_ρ the k-sheeted circle $|z| = e^{2\pi k\rho}$, and $l(\rho) = 1$. On this surface consider the function

(126) $$f(z) = \sqrt[k]{\frac{e^z}{e^z-1}}.$$

23B. This function is single-valued and meromorphic on the covering surface R. Clearly f omits $\zeta = 0$ and the k values $e^{2\pi ji/k}$, $j=0, 1, \cdots, k-1$. The poles are at $z = 2\pi ji$. Let Δ be a small disk containing the point at infinity. Using notations of 21 we have $n_m(\Delta) \sim 2m$. Moreover $e_m \sim 2(k-1)m$ and $\xi \le k-1$. It follows that the number $k+1$ of exceptional values is at most $2+\xi$, i.e., $k-1 \le \xi$. Consequently $\xi = k-1$ and we infer that the nonintegrated form (114) is sharp for every nonnegative integer ξ.

APPENDIX I

BASIC PROPERTIES OF RIEMANN SURFACES

In this appendix we shall derive *ab ovo* some properties of Riemann surfaces to which we have made reference. For the convenience of the reader we list here the headings.

§1. *Characterization of parabolicity.*—1. Capacity function.—2. Harmonic measure.—3. Positive superharmonic functions.—4. Characterization of O_G.—5. Proof.—6. Sets of vanishing capacity.—7. Functional capacity.—8. Equivalence.—9. Polar sets.

§2. *Modular O_G- and O_{AB}-tests.*—10. Modulus.—11. Geometric meaning.—12. Modulus of a regular open set.—13. Modular inequality.—14. O_G-test.—15. Graph.—16. O_{AB}-test.—17. Modular form of the test.—18. Integrated form of the test.—19. Wide exhaustions.

§3. *Directed nets.*—20. Directed sets.—21. Fatou's lemma.—22. Examples.

The grouping of these topics is as follows. In 1 to 5 we enumerate various ways of defining parabolicity and show their equivalence. Point sets of vanishing capacity are studied in 6 to 9. Basic properties of moduli are the topic of 10 to 13, modular O_G-tests of Sario [2] and Noshiro [5] are derived in 14 and 15, and O_{AB}-tests of Pfluger [2], Mori [2], and Kuroda in 16 to 19. Fatou's lemma for directed nets, together with examples, is discussed in 20 to 22.

§1. CHARACTERIZATION OF PARABOLICITY

1. Capacity function. Given an open Riemann surface R and a point $z_0 \in R$ choose an arbitrary but then fixed parametric disk D containing z_0. Let Ω with boundary β_Ω be a regular subregion of R containing $\bar{D}$. For any set E we denote by $H(E)$ the family of harmonic functions on E.

The *capacity function* p_Ω of Ω with singularity at z_0 is defined by the conditions $p_\Omega \in H(\bar{\Omega}-z_0)$,

$$p_\Omega \,|\, \bar{D} = \log|z-z_0| + h(z),$$

and $p_\Omega \,|\, \beta_\Omega = k_\Omega$ (const.), where $h \in H(\bar{D})$ with $h(z_0)=0$. The coefficient of $\log|z-z_0|$ is immaterial; we have sometimes used $1/2\pi$ to obtain unit flux, while here 1 will be more convenient.

The *Green's function* g_Ω of Ω with singularity at z_0 is the function in $H(\bar{\Omega}-z_0)$ with

$$g_\Omega \,|\, \bar{D} = -\log|z-z_0| + \bar{h}(z),$$

where $\bar{h} \in H(\bar{D})$ and $g_\Omega \,|\, \beta_\Omega = 0$. Trivially

$$p_\Omega = k_\Omega - g_\Omega, \tag{1}$$

and $\bar{h}(z_0) = k_\Omega$ is the *Robin constant* of Ω with respect to z_0 and D. It gives the *capacity* $c_{\beta_\Omega} = e^{-k_\Omega}$ of β_Ω with respect to z_0 and D.

For $\Omega \subset \Omega'$ the maximum principle applied to $g_{\Omega'} - g_\Omega$ shows that $\bar{h}(z_0)$ increases with Ω and a fortiori $k_\Omega \le k_{\Omega'}$. Therefore the directed limit

$$k_\beta = \lim_{\Omega \to R} k_\Omega$$

exists; it could be called the Robin constant of R with respect to z_0 and D. It gives the capacity

$$c_\beta = e^{-k_\beta} = \lim_{\Omega \to R} c_{\beta_\Omega}$$

of the ideal boundary β of R with respect to z_0 and D.

An open Riemann surface R is *parabolic* or *hyperbolic* according as $c_\beta = 0$ or $c_\beta > 0$. We include closed surfaces in the class O_G of parabolic surfaces so defined.

It is easily seen that the directed limit

$$g_R(z) = \lim_{\Omega \to R} g_\Omega(z)$$

exists if and only if $R \notin O_G$. We call g_R the Green's function of R with singularity z_0.

2. Harmonic measure. Given an open Riemann surface R choose a regular region Ω_0 of R with boundary β_0 and with connected complement. Let Ω be a regular region of R containing $\bar{\Omega}_0$. The function $u_\Omega \in H(\bar{\Omega} - \Omega_0)$, with $u_\Omega \,|\, \beta_\Omega = 1$, $u_\Omega \,|\, \beta_0 = 0$, is the *harmonic measure* of β_Ω with respect to $\Omega - \bar{\Omega}_0$.

For $\Omega \subset \Omega'$ the maximum principle gives $0 < u_{\Omega'} \le u_\Omega$, and we can define the harmonic measure u_R of β with respect to $R - \bar{\Omega}_0$ as the directed limit

$$u_R(z) = \lim_{\Omega \to R} u_\Omega(z).$$

In the case $u_R \equiv 0$ we say that *the harmonic measure of β vanishes*. We include in this category the harmonic measure of the ideal boundary of a closed surface, since this boundary is empty.

3. Positive superharmonic functions. Consider a positive superharmonic function v defined on $R-\bar{\Omega}_0$ which is continuously extendable to $R-\Omega_0$. If $v \geq m$ (const.) on β_0 implies the same inequality in $R-\Omega_0$ for any v, then we say that *the maximum principle is valid on R*. A closed surface R is a trivial example.

Let $\mathscr{F}=\mathscr{F}(R; z_0)$ be the family of positive superharmonic functions v on R such that $(v+\log|z-z_0|)\,|\,D$ is bounded and superharmonic. We include in the family the constant function ∞. It can be easily seen that $\mathscr{F}$ forms a Perron family and consequently $\inf_{v\in\mathscr{F}} v(z)$ is either identically ∞ or a positive function in $H(R-z_0)$ with a logarithmic singularity at z_0. As a consequence of the maximum principle the latter case can occur if and only if $R \notin O_G$, and then

$$g_R(z) = \inf_{v\in\mathscr{F}} v(z).$$

4. Characterization of O_G. We are now ready to state (cf. Ahlfors [15], Ohtsuka [1]):

Theorem. *The following properties are equivalent:*

(a) *R belongs to the class O_G of surfaces with vanishing capacity of β,*

(b) *the harmonic measure of β vanishes,*

(c) *the maximum principle is valid on R,*

(d) *R does not carry Green's functions,*

(e) *there are no nonconstant positive superharmonic functions on R.*

In particular, (e) assures that the definition of parabolicity does not depend on the choice of z_0 and D. Obviously all closed surfaces enjoy the above properties and we have only to prove the theorem for open surfaces.

5. Proof. Suppose R has property (a) and hence (d). To see that this implies (b) we first note that the minimum m_Ω of g_Ω on β_0 diverges to ∞ as $\Omega \to R \in O_G$. By Green's formula $\int_{\beta_0} g_\Omega\, du_\Omega^* = 2\pi$ and consequently

$$D_\Omega = \int_{\beta_\Omega} u_\Omega du_\Omega^* = \int_{\beta_0} du_\Omega^* \leq \frac{2\pi}{m_\Omega},$$

where D_Ω is the Dirichlet integral of u_Ω over $\Omega-\bar{\Omega}_0$. Therefore $D_\Omega \to 0$ as $\Omega \to R$, and the function u_Ω converges to zero in view of $u_\Omega\,|\,\beta_0=0$. We conclude that $u_R=0$, and condition (b) is satisfied.

Suppose now (b) holds. If $v\,|\,\beta_0 \geq m$, then by the maximum principle $v-m(1-u_\Omega)\geq 0$ on $\bar{\Omega}-\Omega_0$. On letting $\Omega \to R$ we have $v-m\geq 0$ on $R-\bar{\Omega}_0$, and condition (c) is met.

The proof that (c) implies (d) is by contradiction. Assume indeed that there exists a Green's function g_R of R and let $D_\rho=\{z\,|\,z\in R,\, g_R(z)>\rho\}$. For sufficiently large ρ, D_ρ is a relatively compact "disk" about z_0, and $D_\rho \to z_0$ as $\rho\to\infty$. By (c), $g_R \geq \rho$ on $R-D_\rho$. On letting $\rho\to\infty$ we obtain $g_R=\infty$ on $R-z_0$, a contradiction.

We proceed to show that (e) is a consequence of (d). Let v be a nonconstant positive superharmonic function on R. We may assume that $\sup_R v > 1 > \inf_R v$. Since $\min(1, v)$ is again a nonconstant positive superharmonic function on R, we may also suppose that $v = 1$ in a relatively compact disk Ω_0 about some point z_0 in R and that $\inf_R v < 1$. Clearly $v \geq 1 - u_\Omega$ on $\bar{\Omega} - \Omega_0$ and therefore $v \geq 1 - u_R$, whence $u_R > 0$ in $R - \bar{\Omega}_0$.

Take a local parameter z such that $\bar{\Omega}_0 = \{z \mid |z - z_0| \leq 1\} \subset \{z \mid |z - z_0| \leq 2\}$ and set $a = (\min_{|z - z_0| = 2} u_R(z))^{-1}$. Consider the function

$$q(z) = \begin{cases} a \log 2 - \log |z - z_0| & \text{in } \bar{\Omega}_0, \\ a \log 2 - a u_R(z) \log 2 & \text{in } R - \bar{\Omega}_0. \end{cases}$$

It is easy to see that $q \in \mathscr{F}(R; z_0)$ and we infer that g_R exists.

The remaining statement, (a) implied by (e), is immediate. In fact, if $R \notin O_G$, then as remarked in 1, g_R exists and is a nonconstant superharmonic function on R.

The proof of Theorem 4 is herewith complete.

6. Sets of vanishing capacity. By definition a compact set K of the extended z-plane has *logarithmic capacity zero* or simply *capacity zero* if the complement CK belongs to the class O_G.

A general set E in the extended z-plane is said to have capacity zero, or more precisely inner capacity zero, if every compact subset of E has capacity zero.

Let E be a subset of a Riemann surface R. Suppose that $E \cap D$ has capacity zero for every parametric disk D of R. Then we say that E has capacity zero.

7. Functional capacity. Given a Riemann surface R and a positive continuous function k in $R \times R$ in the extended sense (cf.IV.1A) we put for a compact set $K \subset R$

$$I_k(K) = \inf \int k(z_1, z_2)\, d\mu(z_1)\, d\mu(z_2),$$

where μ runs over all Borel measures whose supports are contained in K, and $\mu(K) = 1$. If $I_k(K) = \infty$, then we say that K has *k-capacity zero*. The extension to an arbitrary set E is similar to 6.

Consider two functions k_1 and k_2 such that $k_1 - k_2$ is bounded on $K \times K$. It is readily seen that K has k_1-capacity zero if and only if it has k_2-capacity zero. In particular, *the concept of g_R-capacity zero is equivalent to that of log-capacity zero.*

8. Equivalence. Let K be a compact set in the finite z-plane. Assume that $I_{\log}(K) < \infty$. We can construct a nonconstant positive superharmonic function on CK (cf. Tsuji [19, p. 60]). Thus $CK \notin O_G$ and K has positive capacity.

Conversely suppose $CK \notin O_G$. Then $u_{CK} > 0$. By Lemma IV.4A we conclude that $I_{g_{CD}}(K) < \infty$, where D is a disk with $\bar{D} \subset CK$.

The notion of log-capacity zero is equivalent to that of capacity zero.

9. Polar sets. Let R be a closed Riemann surface, Ω_0 a regular region of R, and K a compact set in $R - \bar{\Omega}_0$. The set K is called a *polar set* if there exists a positive superharmonic function v in $R - \bar{\Omega}_0$ such that $v = \infty$ on K.

In such a case we have by the maximum principle $u_{R-K} = 0$ and therefore $R - K \in O_G$, i.e., K is a set of vanishing capacity.

Conversely assume that $R - K \in O_G$. Let Ω be a regular region in $R - K$ such that $\Omega \supset \bar{\Omega}_0$. We set $u_\Omega = 1$ on $R - \Omega$, where u_Ω refers to Ω_0. Since $u_\Omega(z_1) \to 0$ for a fixed point $z_1 \in R - \bar{\Omega}_0 - K$, there exists an exhaustion $\{\Omega_n\}$ of $R - K$ such that

$$v = \sum_{n=1}^{\infty} u_{\Omega_n}$$

is a positive superharmonic function in $R - \bar{\Omega}_0$. Clearly $v = \infty$ on K and thus K is a polar set. We have shown:

A compact set is a polar set if and only if it has vanishing capacity.

§2. MODULAR O_G- AND O_{AB}-TESTS

10. Modulus. Consider a regular region Ω of a Riemann surface R. We assume that the boundary $\partial\Omega$ of Ω consists of at least two contours and that the contours are divided into two sets α and β. We also allow the degenerate case where the contours are piecewise analytic and not necessarily simple.

Let $u \in H(\bar{\Omega})$ with $u \mid \alpha = 0$, $u \mid \beta = \log \mu$, where $\mu > 1$ is a constant determined by the condition $\int_\alpha du^* = 2\pi$. The number $\log \mu$ is called the *modulus* of the configuration (Ω, α, β) and is denoted by

$$\operatorname{mod}(\Omega, \alpha, \beta) = \operatorname{mod}\Omega = \log \mu.$$

The function u is, by definition, the modulus function.

Sometimes μ itself is called the modulus, and $\log \mu$ is referred to as the logarithmic modulus or harmonic modulus. However, we have used the term modulus for $\log \mu$ throughout the book.

11. Geometric meaning. Let u^* be the (multiple-valued) conjugate harmonic function of u. The analytic mapping given by $\zeta = e^{u + iu^*}$ indicates the geometric meaning of the modulus $\log \mu$. If Ω is doubly connected, then ζ maps Ω conformally onto the annulus with radii 1 and μ, and the modulus is the logarithm of the ratio of the radii of this image.

If Ω is planar but α consists of one contour and β of two contours β_1 and β_2, then we cut Ω along some level line of u^* joining β_1 and β_2. The function ζ maps the resulting doubly connected region onto a radial slit annulus with radii 1 and μ.

Next consider the case where $\partial\Omega$ consists of two contours α and β but the genus of Ω is 1. Cut Ω along a level line of u^* which is a Jordan curve not dividing Ω. The resulting surface is planar and is again mapped by ζ onto a radial slit annulus with radii 1 and μ.

By using both methods of cutting, one for contours and the other for genus, we obtain from an arbitrary Ω a planar region which is mapped by ζ onto a radial slit annulus with radii 1 and μ. We conclude:

The modulus is the logarithm of the ratio of the radii of the image radial slit annulus.

12. Modulus of a regular open set. If Ω consists of a finite number of disjoint regular regions Ω^i, then we call Ω a regular open set. Assume that each Ω^i has more than one contour and that the contours are divided into two sets α^i and β^i. We set $\alpha = \bigcup \alpha^i$ and $\beta = \bigcup \beta^i$. We also allow the case where the contours are piecewise analytic and not necessarily simple. In the same manner as in 10 we define the modulus $\log \mu$ of the configuration (Ω, α, β). By virtue of inequality (3) to be now established we see that

$$(2) \qquad \frac{1}{\mathrm{mod}\,(\Omega, \alpha, \beta)} = \sum_i \frac{1}{\mathrm{mod}\,(\Omega^i, \alpha^i, \beta^i)}.$$

13. Modular inequality. Let Ω be a regular open set, and u_Ω the harmonic measure of β, i.e., $u_\Omega \in H(\bar{\Omega})$ with $u_\Omega \mid \alpha = 0$, and $u_\Omega \mid \beta = 1$. By Green's formula $\int_\alpha du_\Omega^* = D_\Omega(u_\Omega)$ and in view of $(\log \mu) u_\Omega = u$

$$(3) \qquad \mathrm{mod}\,\Omega = \log \mu = \frac{2\pi}{D_\Omega(u_\Omega)}.$$

Let γ be a finite set of disjoint analytic Jordan curves in Ω separating α from β and dividing Ω into two open sets Ω_1 and Ω_2 such that $\alpha \subset \partial\Omega_1$ and $\beta \subset \partial\Omega_2$. Let v_1 be the harmonic measure of γ in Ω_1, and v_2 that of β in Ω_2. We define the function f_λ on $\bar{\Omega}$ by

$$f_\lambda = \begin{cases} \lambda v_1 & \text{on} \quad \alpha \cup \Omega_1 \cup \gamma, \\ (1-\lambda)\left(v_2 + \dfrac{\lambda}{1-\lambda}\right) & \text{on} \quad \gamma \cup \Omega_2 \cup \beta, \end{cases}$$

where λ is a real number between 0 and 1. By Green's formula $D_\Omega(f_\lambda) = D_\Omega(u_\Omega) + D_\Omega(u_\Omega - f_\lambda)$ and therefore

$$D_\Omega(u_\Omega) \le D_\Omega(f_\lambda) = \lambda^2 D_{\Omega_1}(v_1) + (1-\lambda)^2 D_{\Omega_2}(v_2).$$

We choose $\lambda = D_{\Omega_2}(v_2)/(D_{\Omega_1}(v_1) + D_{\Omega_2}(v_2))$ and conclude that

$$\frac{1}{D_\Omega(u_\Omega)} \ge \frac{1}{D_{\Omega_1}(v_1)} + \frac{1}{D_{\Omega_2}(v_2)}.$$

By (3) this is the desired inequality

(4) $$\operatorname{mod}(\Omega, \alpha, \beta) \ge \operatorname{mod}(\Omega_1, \alpha, \gamma) + \operatorname{mod}(\Omega_2, \gamma, \beta).$$

14. O_G-test. Let $\{R_n\}_0^\infty$ be an exhaustion of an open Riemann surface R, with $\log \mu_n = \operatorname{mod}(R_n - \bar{R}_{n-1}, \partial R_{n-1}, \partial R_n)$ for $n = 1, 2, \cdots$. Here and in the sequel we again allow the case where the contours of R_n are piecewise analytic and not necessarily simple. We shall prove the following modular criterion of parabolicity (Sario [2] and Noshiro [5]):

Theorem. *An open Riemann surface R belongs to O_G if and only if there exists an exhaustion of R such that*

(5) $$\sum_{n=1}^{\infty} \log \mu_n = \infty.$$

Proof. Let v_m be the harmonic measure of ∂R_m with respect to $R_m - \bar{R}_0$. Then by (3) and (4)

$$\frac{2\pi}{D_m} \ge \sum_{n=1}^{m} \log \mu_n,$$

where D_m is the Dirichlet integral of v_m over $R_m - \bar{R}_0$. Therefore (5) implies that $D_m \to 0$ as $m \to \infty$. This shows that the harmonic measure of the ideal boundary of R vanishes, i.e., $R \in O_G$.

Conversely assume that $R \in O_G$ and take an arbitrary exhaustion $\{\Omega_n\}_0^\infty$ of R. Fix an integer $m \ge 1$. Let v_n and v_{nm} be the harmonic measures of ∂R_n with respect to $\Omega_n - \bar{\Omega}_0$ and $\Omega_n - \bar{\Omega}_m$, $n > m$, respectively. Since $v_n > v_{nm}$ and $v_n \to 0$ as $n \to \infty$, we have $v_{nm} \to 0$ and therefore by (3)

$$\lim_{n \to \infty} \log \mu_{nm} = \infty,$$

where $\log \mu_{nm} = \operatorname{mod}(\Omega_n - \bar{\Omega}_m, \partial\Omega_m, \partial\Omega_n)$. Thus we can choose a suitable subsequence $\{R_n\}_0^\infty$ of $\{\Omega_n\}$ for which (5) holds.

15. Graph. Take an exhaustion $\{R_n\}_0^\infty$ of an open Riemann surface R. Let u_n and $\log \mu_n$ be the modulus function and the modulus of $(R_n - \bar{R}_{n-1}, \partial R_{n-1}, \partial R_n)$, respectively. The function $u_n + iu_n^*$ maps

$R_n - \bar{R}_{n-1}$ with a finite number of suitable slits conformally onto a slit rectangle $0 < u_n < \log \mu_n$, $0 < u_n^* < 2\pi$. Let

$$(6) \qquad \zeta = \xi + i\eta = u_n + iu_n^* + \sum_0^{n-1} \log \mu_i, \qquad \mu_0 = 1,$$

on $R_n - R_{n-1}$, $n = 1, 2, \cdots$. The function ζ maps $R - \bar{R}_0$ with at most a countable number of suitable slits conformally onto a countably slit strip $0 < \xi < L$, $0 < \eta < 2\pi$ with $L = \sum_1^\infty \log \mu_i$. This strip is the *graph* of R associated with $\{R_n\}$ and L is the *length* of the graph (Noshiro [5]).

In terms of the graph Theorem 14 is restated as follows:

Theorem. *An open Riemann surface R belongs to O_G if and only if there exists an exhaustion such that the length of the graph is infinite.*

16. O_{AB}-test. We turn to modular tests for a given R to belong to the class O_{AB} of Riemann surfaces without nonconstant bounded analytic functions. Let γ_λ be the level line $\xi = \lambda$, where ξ is the real part of the function in (6) and $\lambda \in [0, L)$. For each λ, γ_λ consists of a finite number of piecewise analytic Jordan curves γ_λ^j, $j = 1, \cdots, m(\lambda)$. We consider the variation

$$\Lambda_j(\lambda) = \int_{\gamma_\lambda^j} d\xi^*$$

of ξ^* and set

$$\Lambda(\lambda) = \max_{1 \le j \le m(\lambda)} \Lambda_j(\lambda).$$

Using this maximum variation and the quantity

$$M(\lambda) = \max_{0 \le \rho \le \lambda} m(\rho)$$

we first state the following criterion of Pfluger [2]:

Theorem. *If the maximum variation satisfies the condition*

$$(7) \qquad \limsup_{\lambda \to L} \left(4\pi \int_0^\lambda \frac{d\lambda}{\Lambda(\lambda)} - \log M(\lambda) \right) = \infty,$$

then R belongs to O_{AB}.

Proof. Suppose there exists a nonconstant AB-function f on R; we may assume $|f| < 1$. We denote by $S(\lambda)$ the area of the multisheeted Riemannian image $f(R_0 \cup \{z \mid \xi(z) < \lambda\})$ and by l_λ^j the length of the Riemannian image $f(\gamma_\lambda^j)$, both in the Poincaré metric of the unit disk. Since

$$\Lambda_j(\lambda) = \int_{\gamma_\lambda^j} d\xi^* = \int_{\gamma_\lambda^j} |d\zeta|$$

and

$$l_\lambda^j = \int_{\gamma_\lambda^j} \frac{|df|}{1 - |f|^2} = \int_{\gamma_\lambda^j} \frac{|df/d\zeta|}{1 - |f|^2} |d\zeta|,$$

we obtain by Schwarz's inequality

$$(l_\lambda^j)^2 \le \Lambda_j(\lambda) \int_{\gamma_\lambda^j} \frac{|df/d\zeta|^2}{(1-|f|^2)^2} |d\zeta|$$

for $j=1, \cdots, m(\lambda)$. On summing these inequalities we obtain

$$\sum_1^{m(\lambda)} (l_\lambda^j)^2 \le \Lambda(\lambda) \frac{dS(\lambda)}{d\lambda}.$$

Since the Poincaré metric has constant curvature -4, the corresponding isoperimetric inequality takes the form (Schmidt [1], [2, p. 753, footnote 12])

$$4S_j(\lambda)(\pi+S_j(\lambda)) \le (l_\lambda^j)^2,$$

where $S_j(\lambda)$ is the area of the multisheeted domain bounded by the Riemannian image $f(\gamma_\lambda^j)$. Clearly $S(\lambda) \le \sum_1^{m(\lambda)} S_j(\lambda)$ and therefore

$$S(\lambda)^2 \le \left(\sum_1^{m(\lambda)} S_j(\lambda)\right)^2 \le m(\lambda) \sum_1^{m(\lambda)} S_j(\lambda)^2.$$

This yields

$$4S(\lambda)\left(\pi+\frac{S(\lambda)}{m(\lambda)}\right) \le 4\pi \sum_1^{m(\lambda)} S_j(\lambda)+4 \sum_1^{m(\lambda)} S_j(\lambda)^2$$

$$\le \sum_1^{m(\lambda)} 4S_j(\lambda)(\pi+S_j(\lambda)) \le \sum_1^{m(\lambda)} (l_\lambda^j)^2.$$

Consequently

$$4\pi \frac{d\rho}{\Lambda(\rho)} \le \frac{dS(\rho)}{S(\rho)} - \frac{dS(\rho)}{S(\rho)+\pi M(\lambda)}$$

for $0 \le \rho \le \lambda$, and we obtain by integration

$$4\pi \int_0^\lambda \frac{d\lambda}{\Lambda(\lambda)} \le \log\left(\frac{S(\lambda)}{S(\lambda)+\pi M(\lambda)} \cdot \frac{S(0)+\pi M(\lambda)}{S(0)}\right)$$

$$\le \log \frac{S(0)+\pi M(\lambda)}{S(0)} \le \log\left(M(\lambda)\frac{S(0)+\pi}{S(0)}\right)$$

$$= \log M(\lambda) + \log \frac{S(0)+\pi}{S(0)}.$$

This contradicts (7).

17. Modular form of the test. Let $\{R_n\}_0^\infty$ be an exhaustion of R and let $R_n - \bar{R}_{n-1}$ consist of regions R_{nk}, $k=1, \cdots, k_n$. We denote by $\log \mu_n$ and $\log \mu_{nk}$ the moduli of $(R_n - \bar{R}_{n-1}, \partial R_{n-1}, \partial R_n)$ and $(R_{nk}, (\partial R_{n-1}) \cap (\partial R_{nk}), (\partial R_n) \cap (\partial R_{nk}))$, and by τ_j the jth partial length of the graph, i.e.,

$$\tau_j = \sum_1^j \log \mu_i, \qquad \tau_0 = 0.$$

For the minimum modulus we use the symbol

$$\log \nu_n = \min_{1 \le k \le k_n} \log \mu_{nk}$$

and set

$$N(n) = \max_{1 \le j \le n} k_j.$$

The following modular form of Pfluger's criterion is due to Mori [2]:

Theorem. *If there exists an exhaustion of R with doubly connected surface fragments R_{nk} whose sum of minimum moduli grows so rapidly that*

(8) $$\limsup_{n \to \infty} \left(\sum_{j=1}^{n} \log \nu_j - \tfrac{1}{2} \log N(n) \right) = \infty,$$

then R belongs to O_{AB}.

Proof. Let u_j and u_{jk} be the modulus functions corresponding to $\log \mu_j$ and $\log \mu_{jk}$. Then $u_j = c_{jk} u_{jk}$ in R_{jk} with $c_{jk} = (\log \mu_j)/(\log \mu_{jk})$ and consequently

$$\Lambda_i(\lambda) \le \int_{(\xi=\lambda) \cap R_{jk}} d\xi^* = \int_{(\xi=\lambda) \cap R_{jk}} du_j^* = c_{jk} \int_{(\xi=\lambda) \cap R_{jk}} du_{jk}^* = 2\pi c_{jk}$$

for $\gamma_\lambda^i \subset R_{jk}$. Since k is arbitrary, $\Lambda(\lambda) \le 2\pi c_{jk}$ and a fortiori

$$\Lambda(\lambda) \le 2\pi \frac{\log \mu_j}{\log \nu_j}$$

for $\tau_{j-1} \le \lambda \le \tau_j$. From this and from

$$\int_0^\lambda \frac{d\lambda}{\Lambda(\lambda)} = \sum_1^j \int_{\tau_{i-1}}^{\tau_i} \frac{d\lambda}{\Lambda(\lambda)} + \int_{\tau_j}^\lambda \frac{d\lambda}{\Lambda(\lambda)}$$

it follows for $0 \le \tau_j \le \lambda \le \tau_{j+1}$ that

(9) $$4\pi \int_0^\lambda \frac{d\lambda}{\Lambda(\lambda)} \ge 2 \sum_1^j \log \nu_i + 2 \frac{\log \nu_{j+1}}{\log \mu_{j+1}} (\lambda - \tau_j).$$

In particular

$$4\pi \int_0^{\tau_n} \frac{d\lambda}{\Lambda(\lambda)} \ge 2 \sum_1^n \log \nu_j.$$

Since the R_{nk} are annuli, we have $M(\tau_n) = N(n)$. Therefore

$$4\pi \int_0^{\tau_n} \frac{d\lambda}{\Lambda(\lambda)} - \log M(\tau_n) \ge 2 \left(\sum_1^n \log \nu_j - \tfrac{1}{2} \log N(n) \right),$$

and we infer that (8) implies (7).

18. Integrated form of the test. The following variant of Pfluger's test was established by Kuroda:

Theorem. *If the maximum variation has the property*

$$\limsup_{j\to\infty} \int_{\tau_{j-1}}^{\tau_j} \exp\left(4\pi \int_0^\lambda \frac{d\lambda}{\Lambda(\lambda)}\right) d\lambda = \infty, \tag{10}$$

then R belongs to O_{AB}.

Proof. Suppose there exists a nonconstant analytic function $f = h + ih^*$ with $|f| < 1$ on R. Denote by $D(\lambda)$ the Dirichlet integral of f over $R_0 \cup \{z \mid \xi(z) < \lambda\}$,

$$D(\lambda) = \sum_{j=1}^{m(\lambda)} \int_{\gamma_\lambda^j} h \frac{\partial h}{\partial \xi} d\xi^*.$$

Let h_j be a constant such that $\int_{\gamma_\lambda^j} (h - h_j)\, d\xi^* = 0$. Then the Fourier expansion of $h(\xi^*) - h_j$ takes the form

$$h(\xi^*) - h_j = \sum_{k=1}^{\infty} \left(a_k \cos \frac{2\pi}{\Lambda_j(\lambda)} k\xi^* + b_k \sin \frac{2\pi}{\Lambda_j(\lambda)} k\xi^*\right)$$

for $0 \le \xi^* \le \Lambda_j(\lambda)$. On differentiating both sides with respect to $\xi^* = \eta$ we obtain the Fourier coefficients $\{\Lambda_j(\lambda)^{-1} 2\pi k a_k, \Lambda_j(\lambda)^{-1} 2\pi k b_k\}_1^\infty$ for $\partial h(\xi^*)/\partial\eta$. Therefore

$$\int_{\gamma_\lambda^j} (h - h_j)^2 d\xi^* = \sum_1^\infty (a_k^2 + b_k^2)$$

and

$$\int_{\gamma_\lambda^j} \left(\frac{\partial h}{\partial \eta}\right)^2 d\xi^* = \frac{4\pi^2}{\Lambda_j(\lambda)^2} \sum_1^\infty k^2 (a_k^2 + b_k^2).$$

A comparison of the right-hand sides of these equalities gives

$$\int_{\gamma_\lambda^j} (h - h_j)^2\, d\xi^* \le \frac{\Lambda_j(\lambda)^2}{4\pi^2} \int_{\gamma_\lambda^j} \left(\frac{\partial h}{\partial \eta}\right)^2 d\xi^*.$$

By Schwarz's inequality we have

$$\begin{aligned}
\int_{\gamma_\lambda^j} h \frac{\partial h}{\partial \xi} d\xi^* &= \int_{\gamma_\lambda^j} (h - h_j) \frac{\partial h}{\partial \xi} d\xi^* \\
&\le \frac{\Lambda_j(\lambda)}{2\pi} \sqrt{\int_{\gamma_\lambda^j} \left(\frac{\partial h}{\partial \eta}\right)^2 d\xi^* \int_{\gamma_\lambda^j} \left(\frac{\partial h}{\partial \xi}\right)^2 d\xi^*} \\
&\le \frac{\Lambda_j(\lambda)}{4\pi} \int_{\gamma_\lambda^j} |\text{grad}\, h|^2 d\xi^*.
\end{aligned}$$

It follows that $D(\lambda) \leq (4\pi)^{-1}\Lambda(\lambda)D'(\lambda)$, and by integration we obtain

$$D(0) \exp\left(4\pi \int_0^\lambda \frac{d\lambda}{\Lambda(\lambda)}\right) \leq D(\lambda).$$

On the other hand,

$$\frac{d}{d\lambda} \int_{\gamma_\lambda} h^2 \, d\xi^* = 2 \int_{\gamma_\lambda} h \frac{\partial h}{\partial \xi} \, d\xi^* = 2D(\lambda)$$

and consequently

$$\int_{\tau_{j-1}}^{\tau_j} \exp\left(4\pi \int_0^\lambda \frac{d\lambda}{\Lambda(\lambda)}\right) d\lambda \leq \frac{1}{2D(0)} \left(\int_{\gamma_{\tau_j}} h^2 d\xi^* - \int_{\gamma_{\tau_{j-1}}} h^2 \, d\xi^*\right) \leq \frac{\pi}{D(0)},$$

in violation of (10).

19. Wide exhaustions. Condition (8) can be weakened if we require that the surface fragments R_{jk} are wide enough in the sense that $1/\log \nu_j = O(1)$. On the other hand, we no longer require that the R_{jk} be doubly connected or even planar. The test takes the following form (Kuroda):

Theorem. *If there exists an exhaustion of R such that*

$$\log \nu_j > \delta > 0$$

for $j = 1, 2, \cdots$ and that

$$\limsup_{n \to \infty} \left(\sum_{j=1}^{n} \log \nu_j - \tfrac{1}{2} \log k_n\right) = \infty, \tag{11}$$

then R belongs to O_{AB}.

Proof. Integration of (9) gives

$$J_n = \int_{\tau_{n-1}}^{\tau_n} \exp\left(4\pi \int_0^\lambda \frac{d\lambda}{\Lambda(\lambda)}\right) d\lambda$$

$$\geq \frac{\log \mu_n}{2 \log \nu_n} (\exp(2 \log \nu_n) - 1) \exp\left(2 \sum_1^{n-1} \log \nu_j\right).$$

From (2) it follows that

$$\frac{1}{\log \mu_n} = \sum_1^{k_n} \frac{1}{\log \mu_{nj}},$$

and for this reason $(\log \mu_n)/(\log \nu_n) \geq 1/k_n$. We conclude that

$$J_n \geq \tfrac{1}{2}(1 - \exp(-2\delta)) \exp\left(2 \sum_1^n \log \nu_j - \log k_n\right).$$

This shows that (11) implies (10).

§3. DIRECTED SETS

20. Directed sets. A semiordered set $\Lambda = \{\lambda\}$ with ordering $>$ is, by definition, a *directed set*, if for every pair of elements $\lambda_1, \lambda_2 \in \Lambda$ there exists an element $\lambda_3 \in \Lambda$ such that $\lambda_1 \le \lambda_3$ and $\lambda_2 \le \lambda_3$.

A directed set Λ is called *integral* if it contains a subset Λ_0 with the following properties:

(a) Λ_0 is a sequence $\{\lambda_n\}_1^\infty$ of elements $\lambda_n \in \Lambda$,

(b) $\lambda_m < \lambda_n$ for $m < n$,

(c) for every $\lambda \in \Lambda$ there exists a $\lambda_n \in \Lambda_0$ with $\lambda_n > \lambda$.

An example of directed sets appearing frequently in our book is the set $\{\Omega\}$ of regular subregions Ω of an open Riemann surface R adjacent to a fixed parametric disk R_0 of R. The inclusion relation gives the ordering (cf. I.2A).

This directed set is integral. In fact, the set $\{\Omega_n\}_1^\infty$ of regions $\Omega_n = R_n - \bar{R}_0$, where $\{R_n\}_0^\infty$ is an exhaustion of R, clearly enjoys properties (a), (b), and (c).

A mapping of a directed set $\Lambda = \{\lambda\}$ into a set X is called a *directed net* of elements in X with index set Λ; we usually denote it by $\{x_\lambda\}$ or $\{x_\lambda\}_{\lambda \in \Lambda}$, where x_λ is the image of λ in X.

The fundamental functions $A(k, a)$, $B(k, a)$, and $C(k)$ introduced in I.2 and II.9 can be considered to form directed nets. In fact, these are not functions of real numbers k but functions of regions $\Omega \in \{\Omega\}$ determining $k = k(\Omega)$. Therefore $\{A(k, a)\}$, $\{B(k, a)\}$, and $\{C(k)\}$ are directed nets with index set $\{\Omega\}$.

21. Fatou's lemma. Theorems concerning the interchange of limit and integration are no longer true in general for directed nets because measure theory presupposes the countability property. However if the index set is integral, then such theorems remain valid under certain restrictions. Typical is the following lemma of Fatou:

Theorem. *Let* $\{\varphi_\lambda\}_{\lambda \in \Lambda}$ *be a directed net of functions* φ_λ *on a measure space* (X, μ). *Suppose that*

(a) $\{\varphi_\lambda\}$ *is uniformly bounded from below,*

(b) φ_λ *and* $\inf_{\lambda' > \lambda} \varphi_{\lambda'}$ *are measurable for every* λ,

(c) *the index set* Λ *is integral.*

Then the function $\liminf_\lambda \varphi_\lambda$ *is measurable and*

$$\liminf_{\lambda} \int \varphi_\lambda \, d\mu \ge \int \liminf_{\lambda} \varphi_\lambda \, d\mu. \tag{12}$$

Here $\liminf_\lambda$ stands, as usual, for $\sup_\lambda \inf_{\lambda' > \lambda}$.

Proof. Let $\Lambda_0=\{\lambda_n\}_1^\infty$ be a subset of Λ with properties (a), (b), and (c) in 20. Clearly $\int\varphi_{\lambda''}\,d\mu\geq\int\inf_{\lambda'>\lambda}\varphi_{\lambda'}\,d\mu$ for every $\lambda''>\lambda$; hence

$$\liminf_{\lambda}\int\varphi_\lambda\,d\mu\geq\int\inf_{\lambda'>\lambda}\varphi_{\lambda'}\,d\mu$$

for every λ, and

$$\liminf_{\lambda}\int\varphi_\lambda\,d\mu\geq\sup_{\lambda}\int\inf_{\lambda'>\lambda}\varphi_{\lambda'}\,d\mu. \tag{13}$$

Set $\psi_n=\inf_{\lambda'>\lambda_n}\varphi_{\lambda'}$ and note that $\inf_{\lambda'>\lambda}\varphi_{\lambda'}$ increases with λ. For this reason $\sup_\lambda\int\inf_{\lambda'>\lambda}\varphi_{\lambda'}\,d\mu\geq\int\psi_n d\mu$, and by (13) we conclude that

$$\liminf_{\lambda}\int\varphi_\lambda d\mu\geq\lim_n\int\psi_n d\mu=\int\lim_n\psi_n d\mu. \tag{14}$$

The last equality is assured by the usual Fatou lemma.

By 20(c) there exists an n such that $\psi_n\geq\inf_{\lambda'>\lambda}\varphi_{\lambda'}$ for a given λ; hence $\lim_n\psi_n\geq\inf_{\lambda'>\lambda}\varphi_{\lambda'}$ for every λ, and finally $\lim_n\psi_n\geq\liminf_\lambda\varphi_\lambda$.

On the other hand, $\inf_{\lambda'>\lambda}\varphi_{\lambda'}\geq\psi_n$ for $\lambda>\lambda_n$ and therefore $\liminf_\lambda\varphi_\lambda\geq\psi_n$ for every n. We infer that $\liminf_\lambda\varphi_\lambda=\lim_n\psi_n$ and that $\liminf_\lambda\varphi_\lambda$ is measurable. This together with (14) yields the validity of (12).

22. Examples. First we remark that the function

$$B_k(a)=\int_{\beta_k}s(f(z),a)\,du^*$$

considered in II.12B and a fortiori $B(k,a)$ and $A(k,a)=C(k)-B(k,a)$ are finitely continuous with respect to a on S. This follows from Theorem II.14B. In fact, if $a_0\notin f(\beta_k)$, then the continuity of B_k at a_0 is a direct consequence of the finite joint continuity of s. If $a_0\in f(\beta_k)$, then by II.(79) we have only to establish the continuity of

$$\int_{\beta_k}g_G(f(z),a)\,du^*$$

at a_0, where g_G is the Green's function of a regular region G containing $f(\beta_k)$.

Let $f^{-1}(a_0)\cap\beta_k=\{z_1,\cdots,z_n\}$; let $\beta_{k\nu}$ be a small open subarc of β_k containing z_ν, $\nu=1,\cdots,n$, and set $\beta_k'=\beta_k-\bigcup_\nu\beta_{k\nu}$. Since the function $(z,a)\to g_G(f(z),a)$ behaves logarithmically near (z_ν,a_0),

$$\int_{\beta_{k\nu}}g_G(f(z),a)\,du^*-\int_{\beta_{k\nu}}g_G(f(z),a_0)\,du^*=O(\varepsilon)$$

for a given $\varepsilon>0$ and for (z,a) in the vicinity of (z_ν,a_0), if we choose $\beta_{k\nu}$ sufficiently small. By the continuity of $\int_{\beta_k'}g_G(f(z),a)\,du^*$ at $a=a_0$ we conclude on that of $\int_{\beta_k} g_G(f(z), a)\, du^*$ at a_0.

The continuity of $A(k, a)$ can also be easily deduced from its very definition.

Let μ be a Borel measure on S. The measure ω in II.8A is an example of μ. As a consequence of Theorem 21 we state:

The function $B_k(a)$ *is continuous,* $\inf_{\Omega' \supset \Omega} B_{k'}(a)$ *with* $k' = k(\Omega')$ *is upper semicontinuous, and* $\liminf_{\Omega} B_k(a)$ *with* $k = k(\Omega)$ *is Borel measurable. Moreover*

$$\liminf_{\Omega} \int B_k(a)\, d\mu(a) \geq \int \liminf_{\Omega} B_k(a)\, d\mu(a)$$

with $k = k(\Omega)$.

APPENDIX II

GAUSSIAN MAPPING OF ARBITRARY MINIMAL SURFACES

A smoothly immersed oriented surface R in Euclidean 3-space E^3 is *minimal*, by definition, if its mean curvature vanishes. The natural metric of E^3 induces a conformal structure making R into a Riemann surface. Similarly an oriented unit sphere Σ in E^3 is a Riemann surface, and the radius of Σ parallel to the normal $n(z)$ of R at $z \in R$ gives a conformal mapping $n = n(z)$ of R into Σ. This is the *Gaussian mapping*.

In 1 to 4 of this appendix we shall give an explicit construction of complete minimal surfaces of arbitrary finite or infinite connectivity and genus, smoothly immersed in the Euclidean 3-space.

In 5 to 8 we study the distribution of normals to an arbitrary oriented minimal surface R. Given a point a of the unit sphere in E^3, we ask how frequently the radii parallel to the unit normals on a regular subregion Ω exhausting R touch a, and how close in the mean the parallels to normals on $\partial\Omega$ come to a.

A simple illustration of Theorem 5 is the distribution of normals to a catenoid: the omission of (two) directions is compensated for by a close proximity to those directions by the normals on $\partial\Omega$. For the number of Picard directions in the general case we now have the explicit expression (9) for η in the bound (10).

At the end of the appendix we give the nonintegrated forms of the main theorems. These forms are especially suitable for the study of complete minimal surfaces, on which the natural induced metric is always available to explicitly test regular exhaustibility.

A list of open questions is included.

The result in 1 is due to Osserman [2], and those in 2 to 8 to Klotz-Sario [1], [2].

1. Triple connectivity. Osserman's proof that there exist triply connected minimal surfaces in E^3 is based on the following theorem [2]:

Suppose there exists a Riemann surface F, a meromorphic function f on F, and a harmonic function h on F such that

(a) *the zeros of the differential*

$$\omega = (h_x - ih_y)\, dz$$

coincide in location and multiplicity with the zeros and poles of f on F, and

(b) *for every closed curve C on F,*

$$\int_C \frac{\omega}{f} = \int_C \overline{f\omega}.$$

Then there exists a conformal immersion X of F into E^3 as a minimal surface $X(F)$ on which

$$ds = \frac{1}{2}\left(|f| + \frac{1}{|f|}\right)|\omega|.$$

To construct a triply connected complete minimal surface (of zero genus) let F be the plane $|z| < \infty$ punctured at -1 and 1. The functions

$$f = \frac{1}{z-1} + \frac{\sqrt{2}}{(z-1)^2} + \frac{1}{z+1} + \frac{\sqrt{2}}{(z+1)^2},$$

$$h = \log|z^2-1| - 2\sqrt{2}\,\mathrm{Re}\,\frac{z}{z^2-1}$$

give $\omega = f\,dz$, and condition (a) is satisfied. For a closed curve C in F

$$\int_C \frac{\omega}{f} = \int_C dz = 0,$$

$$\int_C f\omega = \int_C f^2\,dz = 0,$$

as the residues of f^2 at -1, 1 vanish. A conformal immersion $X(F)$ in E^3 thus exists.

We have $ds \geq |dz|/2$ on F, while $|f|^2 \sim 2/|z+1|^4$ at -1 and $|f|^2 \sim 2/|z-1|^4$ at 1. One concludes that the images on $X(F)$ of the paths in F to -1, 1, and ∞ have infinite length, and the surface $X(F)$ is complete.

2. Arbitrary connectivity. We can now construct a complete minimal surface F^c of arbitrary finite or infinite connectivity c and zero genus. In view of the plane and the catenoid we may take $c > 3$.

Make a slit α in the surface F of 1 along the real axis from -1 to 1 and, if $c < \infty$, take $c-2$ copies $F_1, \cdots, F_{c-2}$ of such slit surfaces. Join the copies into one surface F^c by identifying the lower edge of α on F_i for $1 \leq i \leq c-3$ with the upper edge of α on F_{i+1}, and the lower edge of α on F_{c-2} with the upper edge of α on F_1. If $c = \infty$, form F^c by taking infinitely many copies $\cdots F_{-2}, F_{-1}, F_0, F_1, F_2, \cdots$ of the slit surface F and by identifying the lower edge of α on F_i with the upper edge of F_{i+1} for all i. In each case the surface F^c has zero genus and the desired connectivity c.

The natural extension X^c of X to F^c gives a conformal immersion $X^c(F^c)$ of F^c as a complete minimal surface in E^3.

3. Arbitrary genus. If no requirements are made on the number of boundary components, the simplest way to construct a complete minimal surface G^g of given finite or infinite genus g is to use two copies F_j^c, $j=1, 2$, of the surface F^c constructed above, with $c=2g+4$. For $g<\infty$ each F_j^c consists of $2g+2$ copies F_{ij}, $i=1, \cdots, 2g+2$, of F slit along α. We cut each F_j^c along $g+1$ slits β_{kj}, $k=1, \cdots, g+1$, each consisting of the lower half of the imaginary axis on $F_{2k-1,j}$ and the upper half of the imaginary axis on $F_{2k,j}$. The desired surface G^g is obtained by identifying the left edge of β_{k1} with the right edge of β_{k2}, and vice versa. If $g=\infty$, both copies of F_j^∞ consist of infinitely many duplicates F_{ij}, $i=\cdots -2, -1, 0, 1, 2, \cdots$, and the same construction gives G^∞. In both cases the natural extension of X to G^g gives a conformal immersion of G^g into E^3 as a complete minimal surface.

4. Arbitrary genus and connectivity. For short we shall refer to the number of components of the ideal boundary (see, e.g., Ahlfors-Sario, [1, p. 67]) as the connectivity of the surface even if the genus is positive. The connectivity of G^g constructed above is uniquely determined by the genus g and increases with it. To construct a surface of arbitrary connectivity (≥ 4) and genus g we shall first form a surface $H^{4,g}$ of connectivity 4 and genus g.

If $g<\infty$, let F_m^4, $m=1, \cdots, g+1$, be duplicates of F^4, each consisting of two copies F_{im}, $i=1, 2$, of F joined along α. On each F_m^4 make two slits on the real axis: γ_{1m} on F_{1m} from $-\infty$ to -1 and γ_{2m} on F_{2m} from 1 to ∞. For each i identify the lower edge of γ_{im}, $1 \leq m \leq g$, with the upper edge of $\gamma_{i,m+1}$, and the lower edge of $\gamma_{i,g+1}$ with the upper edge of γ_{i1}. If $g=\infty$, we use infinitely many copies F_m^4, $m=\cdots, -2, -1, 0, 1, 2, \cdots$, and identify for each m and i the lower edge of γ_{im} with the upper edge of $\gamma_{i,m+1}$.

The resulting surface $H^{4,g}$ continues to have 4 boundary components. The genus g is, by definition, the largest number of disjoint singular 1-cycles δ_n on $H^{4,g}$ whose union has a connected complement. On each F_m^4, $m=1, \cdots, g$, choose for δ_m the simple closed curve consisting of the circle $|z+1|=1$ on F_{1m} and the circle $|z-1|=1$ on F_{2m}. The curve has the shape of a horizontal figure 8, with the two loops, including the points at $z=0$, lying on different sheets of F_m^4.

A surface $H^{c,g}$ of any connectivity $c \geq 4$ and arbitrary genus g is obtained from $H^{4,g}$ by replacing the copy F_{g+1}^4 in it by F^{c+4} slit along α. Here the slit $\gamma_{1,g+1}$ is made in the first copy F_1 of the $c+2$ duplicates of F that constitute F^{c+4}. The slit $\gamma_{2,g+1}$ is similarly taken in the second copy F_2.

We have arrived at the following result (Klotz-Sario [1]):

Theorem. *There exist complete minimal surfaces, smoothly immersed in E^3, of any connectivity c and any genus g.*

More accurately, for $g=0$, c can be arbitrary, while for $g>0$ we can prescribe any $c \geq 4$.

5. Gaussian mapping. Let Σ be the unit sphere in E^3, intersecting the complex ζ-plane along the circle $|\zeta|=1$, and oriented by choosing its inner normal. To study the distribution of normals to an arbitrary oriented minimal surface R we apply the theory of complex analytic mappings to the Gaussian mapping $n=n(z)$ of R into Σ, both viewed as Riemann surfaces.

As in I.1 and I.9 we use ζ as the local parameter of Σ, and retain the meanings of t_0, s_0, and

$$s(\zeta, a) = \log \frac{(1+|\zeta|^2)(1+|a|^2)}{|\zeta-a|^2}. \tag{1}$$

The mass element

$$d\omega = \frac{4}{(1+|\zeta|^2)^2}\, dS \tag{2}$$

is now the area element of Σ.

In terms of R_h of I.2A we let $\nu(h, a)$ be the number of times, with multiplicity, the normal n to R takes a given direction $a \in \Sigma$ in R. The A-function

$$A(h, a) = 4\pi \int_0^h \nu(h, a)\, dh \tag{3}$$

again reflects the frequency of a-points in R; for a Picard direction it vanishes identically.

We use u, Ω_h, and β_h of I.2A and denote by (n, a) the central angle between n and a. Since $s(\zeta, a) = -2\log(2\sin((\zeta, a)/2))$, the B-function takes the form

$$B(h, a) = -2 \int_{\beta_h - \beta_0} \log \sin \frac{(n, a)}{2}\, du^*. \tag{4}$$

It is the mean proximity of n on $\beta_h - \beta_0$ to a in terms of $s(\zeta, a)$.

The C-function is simply

$$C(h) = -\int_0^h K(h)\, dh, \tag{5}$$

where $K(h)$ is the total curvature $-\int_{R_h} d\omega(n(z))$ of R_h.

Theorem. *For the Gaussian mapping of an arbitrary minimal surface*

$$A(k, a) + B(k, a) = C(k). \tag{6}$$

The meaning of the theorem as discussed in I.2E is particularly fascinating in view of the present concrete significance of the A-, B-, and C-functions. We again refer to the illuminating simple example of the catenoid and the two directions omitted by n.

6. Picard directions. We ask in general how many directions can the normals omit? Given a function φ of $h \in [0, k]$ we again let φ_0 stand for φ in the purely notational formula

$$\varphi_i(h) = \int_0^h \varphi_{i-1}(h)\, dh,$$

$i = 1, 2, 3$. It is the use of these multiply integrated quantities that permits the extension of the value distribution theory to the Gaussian mapping of *arbitrary* minimal surfaces. We denote by $e(h)$ the Euler characteristic of Ω_h and state Theorem I.9B in the present setup:

For any regular subregion Ω of an arbitrary minimal surface R, complete or not,

$$(7) \qquad (2-q)K_3(k) < \sum_1^q A_2(k, a_i) - A_2(k, n') + 4\pi e_3(k) + O(k^3 + k^2 \log(-K_1(k))).$$

Again the meaning of the theorem is that only few directions a_i can remain sparsely covered, for the sum of the counting functions A_2 must in essence dominate $(2-q)K_3$.

For minimal surfaces satisfying the nondegeneracy conditions (I.6C) $k/K_1(k) \to 0$ and

$$(8) \qquad \lim_{R_k \to R} \frac{\log(-K_1(k))}{K_1(ck)} = 0,$$

where c is a constant in $(0, 1)$, we have in terms of the Euler index

$$(9) \qquad \eta = \liminf_{R_k \to R} \frac{4\pi e_3(k)}{-K_3(k)}:$$

Theorem. *The number P of Picard directions has the bound*

$$(10) \qquad P \leq 2 + \eta.$$

More generally,

$$(11) \qquad (\sum \alpha) + \beta \leq 2 + \eta,$$

where α is the defect

$$\alpha(a) = 1 - \limsup_{R_k \to R} \frac{A_2(k, a)}{-K_3(k)},$$

the sum $\sum \alpha$ is taken over any countable set $\{a\}$, and β is the ramification index

$$\beta = \liminf_{R_k \to R} \frac{A_2(k, n')}{-K_3(k)}.$$

Compared with the general value distribution theory the above theorem gains in interest by the explicit nature of the mapping.

If R is an R_p-surface (I.12), then it can again be exhausted by regions $p^{-1}(-\infty, k)$ and the directed limits above can be replaced by ordinary limits. We recall (IV.§1) that, e.g., all parabolic Riemann surfaces are R_p-surfaces.

7. Islands and peninsulas. In this section we shall apply to minimal surfaces the nonintegrated forms of the main theorems (VI.§3 and §5).

Let n again be the Gaussian mapping into Σ of an arbitrary minimal surface R, complete or not. Denote by $K(\Omega)$ the total curvature of a regular region Ω of R. Given a subregion Δ of Σ, its inverse image in Ω consists of two kinds of components, *islands* D_i and *peninsulas* D_p, according as they are relatively compact with respect to Ω or not. Let $m(\Delta)$ be the sum of the total curvatures of the D_i divided by the negative of the spherical area of Δ. Define $\mu(\Delta)$ similarly for $\{D_p\}$. Let L be the "spherical" length of $\partial\Omega$.

For every regular subregion of an arbitrary minimal surface $R \subset E^3$ and for any subregion Δ of the unit sphere Σ the Gaussian mapping gives

$$(12) \qquad 4\pi(m(\Delta)+\mu(\Delta)) = -K(\Omega)+O(L).$$

For minimal surfaces R with L negligible compared with K (see below) this is again a remarkable symmetry: the $(m+\mu)$-affinity is the same for all $\Delta \subset \Sigma$.

We set $e^+(\Omega) = \max\,(e(\Omega), 0)$, where $e(\Omega)$ is the Euler characteristic of Ω.

For disjoint regions $\Delta_1, \cdots, \Delta_q$ of Σ we have

$$(13) \qquad (2-q)K < 4\pi\Big(\sum_\nu m(\Delta_\nu)+e^+(\Omega)\Big)+O(L).$$

For greater accuracy the sum $4\pi \sum_\nu b(\Delta_\nu)$ for the orders b of the branch points above Δ_ν can be subtracted on the right.

The theorem again shows that only few regions Δ_ν can be sparsely covered. For a more precise meaning we now ask: When is L negligible compared with K?

8. Regular exhaustions. We consider a complete minimal surface R. Let $z_0 \in R \subset E^3$ be given and let $\rho(z, z_0)$ be the distance along R from $z \in R$ to z_0 in the Euclidean metric $d\rho$ of E^3. Set

$$\beta_\rho = \{z \mid z \in R,\ \rho(z, z_0) = \rho\},$$
$$\Omega_\rho = \{z \mid z \in R,\ \rho(z, z_0) < \rho\},$$

$0 < \rho < \infty$. For $\rho \to \infty$, Ω_ρ exhausts R.

Let $L(\rho)$, $K(\rho)$ be the quantities L, K for β_ρ, Ω_ρ, respectively. The exhaustion $\Omega_\rho \to R$ is by definition regular if

$$(14) \qquad \liminf_{\rho\to\infty} \frac{L(\rho)}{-K(\rho)} = 0.$$

Denote by $l(\rho)$ the length of β_ρ in E^3. The test VI.18B reads in the present explicit case:

The exhaustion is regular if the negative of the total curvature $K(\rho)$ *increases so rapidly that*

$$\liminf_{\rho \to \infty} \left(K(\rho) \int_\rho^\infty \frac{d\rho}{l(\rho)} \right) = -\infty. \tag{15}$$

We let

$$\xi = \limsup_{\rho \to \infty} \frac{4\pi\, e^+(\rho)}{-K(\rho)} \tag{16}$$

and recall (VI.10D) that the regions Δ_ν can be replaced by points a_ν.

Theorem. *For every regularly exhaustible complete minimal surface the number* P *of Picard directions has the bound*

$$P \leq 2 + \xi. \tag{17}$$

The simplest example is again the catenoid, for which $L \to 0$, $K \to -4\pi$, $L/K \to 0$, $e^+ = 0$, and $\xi = 0$. It follows that $P \leq 2$, and we know that the bound is taken.

Remark. In accordance with the plan of this book we restricted our discussion of Gaussian mappings to applications of the main theorems in our general theory. For other aspects we refer the reader to Osserman [2] to [4], Voss [1], and other literature on Gaussian mappings listed in the Bibliography.

9. Open questions. We close by suggesting some further problems that appear significant in the study of minimal surfaces, known and new, complete or not.

(a) What can be said about specific minimal surfaces in the light of the $(A+B)$-affinity (6)?

(b) Are there complete minimal surfaces whose normals omit a given number of directions?

(c) Are there minimal surfaces with a given $\eta \geq 0$ of (9)?

(d) Which surfaces are nondegenerate in the sense of (8) and thus have at most $2+\eta$ omitted directions?

(e) Is the bound $2+\eta$ sharp in the sense that for any integer $\eta \geq 0$ there are minimal surfaces with this η and with $2+\eta$ omitted directions?

(f) Which complete minimal surfaces are regularly exhaustible in the sense of (14) and thus satisfy (17)?

(g) Questions (c) and (e) for complete minimal surfaces with η replaced by ξ.

BIBLIOGRAPHY

AHLFORS, L. V.

[1] *Beiträge zur Theorie der meromorphen Funktionen.* 7. Scand. Congr. Math. Oslo 1929, pp. 84–88.

[2] *Sur quelques propriétés des fonctions méromorphes.* C.R. Acad. Sci. Paris 190 (1930), 720–722.

[3] *Ein Satz von Henri Cartan und seine Anwendung auf die Theorie der meromorphen Funktionen.* Soc. Sci. Fenn. Comment. Phys.-Math. 5 No. 16 (1930), 19 pp.

[4] *Ein Satz über die charakteristische Funktion und den Maximalmodul einer meromorphen Funktion.* Ibid. 6 No. 9 (1931), 4 pp.

[5] *Sur une généralisation du théorème de Picard.* C.R. Acad. Sci. Paris 194 (1932), 245–247.

[6] *Sur les fonctions inverses des fonctions méromorphes.* Ibid. 194 (1932), 1145–1147.

[7] *Eine Verallgemeinerung des Picardschen Satzes.* Internat. Congr. Math. Zürich 1932, 2, pp. 44–45.

[8] *Über eine in der neueren Wertverteilungstheorie betrachtete Klasse transzendenter Funktionen.* Acta Math. 58 (1932), 375–406.

[9] *Sur les domaines dans lequels une fonction méromorphe prend des valeurs appartenant à une région donnée.* Acta Soc. Sci. Fenn. (2) A 2 No. 2 (1933), 17 pp.

[10] *Über eine Methode in der Theorie der meromorphen Funktionen.* Soc. Sci. Fenn. Comment. Phys.-Math. 8 No. 10 (1935), 14 pp.

[11] *Zur Theorie der Überlagerungsflächen.* Acta Math. 65 (1935), 157–194.

[12] *Geometrie der Riemannschen Flächen.* Internat. Congr. Math. Oslo 1936, 1, pp. 239–248.

[13] *Über die Anwendung differentialgeometrischer Methoden zur Untersuchung von Überlagerungsflächen.* Acta Soc. Sci. Fenn. (2) A 2 No. 6 (1937), 17 pp.

[14] *The theory of meromorphic curves.* Ibid. A 3 No. 4 (1941), 31 pp.

[15] *On the characterization of hyperbolic Riemann surfaces.* Ann. Acad. Sci. Fenn. A I No. 125 (1952), 5 pp.

AHLFORS, L. V.; BEURLING, A.

[1] *Conformal invariants and function-theoretic null-sets.* Acta Math. 83 (1950), 101–129.

AHLFORS, L. V.; SARIO, L.

[1] *Riemann surfaces.* Princeton Univ. Press, Princeton, N.J., 1960. 382 pp.

AHMAD, M.

[1] *A note on a theorem of Borel.* Math. Student 25 (1957), 5–9.

[2] *On exceptional values of entire and meromorphic functions.* Compositio Math. 13 (1958), 150–158.

ÅLANDER, M.

[1] *Über eine Eigenschaft einer meromorphen Funktion innerhalb einer Niveaulinie.* Nordisk Mat. Tidskr. 8 (1926), 52–62.

[2] *Sur une propriété des fonctions méromorphes à l'intérieur d'une ligne de module constant.* C.R. Acad. Sci. Paris 184 (1927), 1411–1413.

ALENICYN, Yu. E.

[1] *An extension of the principle of subordination to multiply connected regions.* Dokl. Akad. Nauk SSSR 126 (1959), 231–234. (Russian.)

ANASTASSIADIS, J.

[1] *Sur les valeurs exceptionnelles des fonctions entières et méromorphes d'ordre fini.* C.R. Acad. Sci. Paris 210 (1940), 204–206.

ANDREIAN CAZACU, C.

[1] *Sur la théorie de R. Nevanlinna.* Acad. R. P. Romiñe. Bul. Şti. Secţ. Şti. Mat. Fiz. 6 (1954), 271–296.

[2] *Suprafeţe Riemanniene parţial regulat exhaustible.* Analele Univ. Bucuresti 11 (1962), 125–154.

ARIMA, K.

[1] *On uniformizing functions.* Kōdai Math. Sem. Rep. (1950), 81–83.

[2] *On a meromorphic function in the unit circle whose Nevanlinna's characteristic function is bounded.* Ibid. (1950), 94–95.

BADESCO, R.

[1] *Sur la distribution des valeurs d'une fonction holomorphe ou méromorphe.* C.R. Acad. Sci. Paris 190 (1930), 911–913.

BAGEMIHL, F.

[1] *Some identity and uniqueness theorems for normal meromorphic functions.* Ann. Acad. Sci. Fenn. Ser. A I No. 299 (1961), 6 pp.

BAGEMIHL, F.; SEIDEL, W.

[1] *Sequential and continuous limits of meromorphic functions.* Ann. Acad. Sci. Fenn. Ser. A I No. 280 (1960), 17 pp.

[2] *Behavior of meromorphic functions on boundary paths, with applications to normal functions.* Arch. Math. 11 (1960), 263–269.

BEHNKE, H.; STEIN, K.

[1] *Entwicklung analytischer Funktionen auf Riemannschen Flächen.* Math. Ann. 120 (1949), 430–461.

BELINSKIĬ, P. P.; GOL'DBERG, A. A.

[1] *Application of a theorem on conformal mappings to questions of invariance of defects of meromorphic functions.* Ukrain. Mat. Ž. 6 (1954), 263–269. (Russian.)

BERGMAN, S.

[1] *Über meromorphe Funktionen von zwei komplexen Veränderlichen.* Compositio Math. 6 (1939), 305–335.

[2] *On value distribution of meromorphic functions of two complex variables.* Studies in mathematical analysis and related topics. Stanford Univ. Press, Stanford, Calif., 1962, pp. 25–37.

[3] *On meromorphic functions of several complex variables.* Abstract of short communications p. 63. Internat. Congr. Math. Stockholm 1962.

BERNSTEIN, V.

[1] *A propos d'une formule de MM.F. et R. Nevanlinna relative aux fonctions méromorphe dans un secteur.* C.R. Acad. Sci. Paris 186 (1928), 1264–1266.

BEURLING, A.

[1] Cf. AHLFORS, L. V.; BEURLING, A. [1].

BIERNACKI, M.

[1] *Sur les directions de Borel des fonctions méromorphes.* C.R. Acad. Sci. Paris 189 (1929), 21–23.

[2] *Sur les directions de Borel des fonctions méromorphes.* Acta Math. 56 (1931), 197–204.

[3] *Sur la caractéristique T(f) des fonctions méromorphes dans un cercle.* Ann. Univ. Mariae Curie-Skłodowska, Sect. A. 9 (1957), 99–125.

BLANC, C.

[1] *Classification des singularités des fonctions inverses des fonctions méromorphes.* C.R. Acad. Sci. Paris 202 (1936), 459–461.

[2] *Les surfaces de Riemann des fonctions méromorphes. I, II.* Comment. Math. Helv. 9 (1937), 193–216; 335–368.

[3] *Une interprétation élémentaire des théorèmes fondamentaux de M. Nevanlinna.* Ibid. 12 (1939–40), 153–163.

BLOCH, A.

[1] *Quelques théorèmes sur les fonctions entières et méromorphes d'une variable.* C.R. Acad. Sci. Paris 181 (1925), 1123–1125.

[2] *Sur la non-uniformisabilité par les fonctions méromorphes des variétés algébriques les plus genérâles.* Ibid. 181 (1925), 276–278.

[3] *La conception actuelle de la théorie des fonctions entières et méromorphes.* Enseignement Math. 25 (1926), 83–103.

[4] *Quelques théorèmes sur les fonctions entières et méromorphes d'une variable.* C.R. Acad. Sci. Paris 182 (1926), 367–369.

[5] *Les fonctions holomorphes et méromorphes dans le cercle-unité.* Gauthier-Villars, Paris, 1926. 61 pp.

[6] *Les théorèmes de M. Valiron sur les fonctions entières et la théorie de l'uniformisation.* Ann. Fac. Sci. Univ. Toulouse (3) 17 (1926), 1–22.

BOHR, H.

[1] *Zum Picardschen Satz.* Mat. Tidsskr. B. (1940), 1–6.

BOHR, H.; LANDAU, R.

[1] *Über das Verhalten von $\zeta(s)$ und $\zeta_k(s)$ in der Nähe der Geraden $\sigma=1$.* Nachr. Akad. Wiss. Göttingen Math.-Phys. Kl. II (1910), 303–330.

BOREL, E.

[1] *Démonstration élémentaire d'un théorème de M. Picard sur les fonctions entières.* C.R. Acad. Sci. Paris 122 (1896), 1045–1048.

[2] *Sur les zéros des fonctions entières.* Acta Math. 20 (1897), 357–396.

[3] *Leçons sur les fonctions méromorphes.* Gauthier-Villars, Paris, 1903. 119 pp.

[4] *Leçons sur les fonctions entières.* Gauthier-Villars, Paris, 1921. 162 pp.

BOSE, S. K.

[1] *A note concerning some properties of the maximum function of a meromorphic function.* Math. Z. 66 (1957), 487–489.

BOUTROUX, P.

[1] *Sur quelques propriétés des fonctions entières.* Acta Math. 28 (1904), 97–224.

BUREAU, F.

[1] *Mémoire sur les fonctions uniformes à point singulier essential isolé.* Mém. Liège (3) 17 No. 3 (1932), 52 pp.

CALUGARÉANO, G.

[1] *Sur la détermination des valeurs exceptionnelles des fonctions entières et méromorphes d'ordre fini.* C.R. Acad. Sci. Paris 188 (1929), 37–39.

[2] *Sur la détermination des valeurs exceptionnelles des fonctions entières et méromorphes de genre fini.* Bull. Sci. Math. (2) 54 (1930), 17–32.

[3] *Une généralisation du théorème de M. Borel sur les fonctions méromorphes.* C.R. Acad. Sci. Paris 192 (1931), 329–330.

[4] *Sur un complément au théorème de M. Borel.* Mathematica 6 (1932), 25–30. Bull. Cluj 6 (1932), 299–304.

[5] *Sur les valeurs exceptionnelles, au sens de M. Picard et de M. Nevanlinna, des fonctions méromorphes.* C.R. Acad. Sci. Paris 195 (1932), 22–23.

[6] *Sur le théorème de M. Borel dans le cas des fonctions méromorphes d'ordre infini.* Mathematica 7 (1933), 61–69. Bull. Cluj 7 (1933), 174–182.

CARATHÉODORY, C.

[1] *Sur quelques généralisations du théorème de M. Picard.* C.R. Acad. Sci. Paris 141 (1905), 1213–1215.

[2] *Über eine Verallgemeinerung der Picardschen Sätze.* S.-B. Deutsch. Akad. Wiss., Berlin Kl. Math. Phys. Tech. (1920), 202–209.

CARLESON, L.

[1] *On a class of meromorphic functions and its associated exceptional sets.* Doctoral dissertation, Univ. Uppsala, 1950. 79 pp.

[2] *A remark on Picard's theorem.* Bull. Amer. Math. Soc. 67 (1961), 142–144.

CARTAN, H.

[1] *Sur quelques théorèmes de M. R. Nevanlinna.* C.R. Acad. Sci. Paris 185 (1927), 1253–1254.

[2] *Un nouveau théorème d'unicité relatif aux fonctions méromorphes.* Ibid. 188 (1929), 301–303.

[3] *Sur la croissance des fonctions méromorphes d'une ou de plusieurs variables complexes.* Ibid. 188 (1929), 1374–1376.

[4] *Sur la fonction de croissance attachée à une fonction méromorphe de deux variables, et ses applications aux fonctions méromorphes d'une variable.* Ibid. 189 (1929), 521–523.

[5] *Sur la derivée par rapport à log r de la fonction de croissance* $T(r; f)$. Ibid. 189 (1929), 625–627.

[6] *Sur les zéros des combinaisons linéaires de p fonctions entières données.* Ibid. 189 (1929), 727–729.

[7] *Sur les valeurs exceptionnelles d'une fonction méromorphe dans tout le plan.* Ibid. 190 (1930), 1003–1005.

CHERN, S.-S.

[1] *Complex analytic mappings of Riemann surfaces. I.* Amer. J. Math. 82 (1960), 323–337.

[2] *The integrated form of the first main theorem for complex analytic mappings in several complex variables.* Ann. of Math. (2) 71 (1960), 536–551.

CHUANG, C.-T.

[1] *Note on the distribution of the values of meromorphic functions of infinite order.* Sci. Rep. Nat. Tsing Hua Univ. A, Peiping, 3 (1936), 457–459.

[2] *Quelques théorèmes sur les directions de Julia et de Borel des fonctions méromorphes.* C.R. Acad. Sci. Paris 204 (1937), 404–405.

[3] *Un théorème relatif aux directions de Borel des fonctions méromorphes d'ordre fini.* Ibid. 204 (1937), 951–952.

[4] *On the distribution of the values of meromorphic functions of infinite order.* J. Chinese Math. Soc. 2 (1937), 21–39.

[5] *Sur la comparaison de la croissance d'une fonction méromorphe et de celle de sa dérivée.* Bull. Sci. Math. (2) 75 (1951), 171–190.

[6] *Une généralisation d'une inégalité de Nevanlinna.* Sci. Sinica 13 (1964), 887–895.

CLUNIE, J.

[1] *The derivative of a meromorphic function.* Proc. Amer. Math. Soc. 7 (1956), 227–229.

[2] *On functions meromorphic in the unit circle.* J. London Math. Soc. 32 (1957), 65–67.

[3] *On integral and meromorphic functions.* Ibid. 37 (1962), 17–27.

COLLINGWOOD, E. F.

[1] *Sur quelques théorèmes de R. Nevanlinna.* C.R. Acad. Sci. Paris 179 (1924), 955–957.

[2] *Sur une théorème de M. Valiron.* Ibid. 182 (1926), 40–42.

[3] *On meromorphic and integral functions.* J. London Math. Soc. 5 (1930), 4–7.

[4] *Sur certaines ensembles définis pour les fonctions méromorphes.* C.R. Acad. Sci. Paris 227 (1948), 615–617.

COLLINGWOOD, E. F.

[5] *Une inégalité dans la théorie des fonctions méromorphes.* Ibid. 227 (1948), 709–711.

[6] *Inégalités relatives à la distribution des valeurs d'une fonction méromorphe dans le plan fini.* Ibid. 227 (1948), 749–751.

[7] *Inégalités relatives à la distribution des valeurs d'une fonction méromorphe dans le cercle unité.* Ibid. 227 (1948), 813–815.

[8] *Exceptional values of meromorphic functions.* Trans. Amer. Math. Soc. 66 (1949), 308–346.

[9] *Conditions suffisantes pour l'inversion de la seconde inégalité fondamentale de la théorie des fonctions méromorphes.* C.R. Acad. Sci. Paris 235 (1952), 1182–1184.

[10] *Relation entre la distribution des valeurs multiples d'une fonction méromorphe et la ramification de sa surface de Riemann.* Ibid. 235 (1952), 1267–1270.

[11] *Sufficient conditions for reversal of the second fundamental inequality for meromorphic functions.* J. Analyse Math. 2 (1952), 29–50.

COLLINGWOOD, E. F.; LOHWATER, A. J.

[1] *Inégalités relatives aux défauts d'une fonction méromorphe dans le cercle-unité.* C.R. Acad. Sci. Paris 242 (1956), 1255–1257.

COLLINGWOOD, E. F.; PIRANIAN, G.

[1] *Tsuji functions with segments of Julia.* Math. Z. 84 (1964), 246–253.

CONSTANTINESCU, C.

[1] *Quelques applications du principe de la métrique hyperbolique.* C.R. Acad. Sci. Paris 242 (1956), 3035–3038.

[2] *Über eine Klasse meromorpher Funktionen, die höchstens einen defekten Wert besitzen können.* Bull. Math. Soc. Sci. Math. Phys. R. P. Roumaine (N.S.) 1 (49) (1957), 131–140.

[3] *Über die defekten Werte der meromorphen Funktionen deren charakteristische Funktion sehr langsam wächst.* Compositio Math. 13 (1958), 129–147.

CONSTANTINESCU, C.; CORNEA, A.

[1] *Analytische Abbildungen Riemannscher Flächen.* Rev. Math. Pures Appl. 8 (1963), 67–72.

[2] *Ideale Ränder Riemannscher Flächen.* Springer-Verlag, Berlin-Göttingen-Heidelberg, 1963. 244 pp.

DELANGE, H.

[1] *Sur les suites de fonctions méromorphes d'ordre borné à zéros et pôles réels et négatifs.* C.R. Acad. Sci. Paris 221 (1945), 741–743.

[2] *Sur certaines fonctions méromorphes.* Ibid. 222 (1946), 40–42.

DINGHAS, A.

[1] *Einige Sätze und Formeln aus der Theorie der meromorphen und ganzen Funktionen.* Math. Ann. 110 (1934), 284–311.

[2] *Zur Theorie der meromorphen Funktionen in einem Winkelraum.* S.-B. Deutsch. Akad. Wiss. Berlin. Kl. Math. Phys. Tech. (1935), 576–596.

[3] *Sur un théorème de Carleman et sur un théorème de Carlson-Nevanlinna.* Bull. Soc. Math. France 64 (1936), 78–86.

[4] *Beiträge zur Theorie der meromorphen Funktionen.* Schr. Math. Sem. Inst. Angew. Math. Univ. Berlin 3 (1936), 67–92.

[5] *Bemerkungen zur Ahlforsschen Methode in der Theorie der meromorphen Funktionen. I.* Compositio Math. 5 (1937), 107–118.

[6] *Über eine Eigenschaft der Charakteristik von meromorphen Funktionen in einer Halbebene.* Math. Z. 44 (1938), 354–361.

[7] *Eine Bemerkung zur Ahlforsschen Theorie der Überlagerungsflächen.* Ibid. 44 (1938), 568–572.

[8] *Eine Verallgemeinerung des Picard-Borelschen Satzes.* Ibid. 44 (1938), 573–579.

[9] *Zur Werteverteilung einer Klasse transzendenter Funktionen.* Ibid. 45 (1939), 507–510.

DINGHAS, A.

[10] *Über Ausnahmegebiete meromorpher Funktionen.* Ibid. 45 (1939), 20–24.

[11] *Zur Invarianz der Shimizu-Ahlforsschen Charakteristik.* Ibid. 45 (1939), 25–28.

[12] *Zur Abschätzung der a-Stellen ganzer transzendenter Funktionen mit Hilfe der Shimizu-Ahlforsschen Charakteristik.* Math. Ann. 120 (1949), 581–584.

[13] *Zu Nevanlinna's zweitem Fundamentalsatz in der Theorie der meromorphen Funktionen.* Ann. Acad. Sci. Fenn. Ser. A I No. 151 (1953), 8 pp.

[14] *Zum Verhalten eindeutiger analytischer Funktionen in der Umgebung einer wesentlichen isolierten Singularität.* Math. Z. 66 (1957), 389–408.

[15] *Über den Picard-Borelschen Satz und die Nevanlinnaschen Ungleichungen.* J. Math. Pures Appl. (9) 42 (1963), 223–251.

DUFRESNOY, J.

[1] *Sur les valeurs exceptionnelles des fonctions méromorphes voisines d'une fonction méromorphe donnée.* C.R. Acad. Sci. Paris 208 (1939), 255–257.

[2] *Sur les fonctions méromorphes dans un angle.* Ibid. 208 (1939), 718–720.

[3] *Sur une propriété de la fonction de croissance $T(r)$ d'un système de fonctions holomorphes.* Ibid. 211 (1940), 536–538.

[4] *Sur les théorèmes fondamentaux de la théorie des courbes méromorphes.* Ibid. 211 (1940), 628–631.

[5] *Sur la théorie d'Ahlfors des surfaces de Riemann.* Ibid. 212 (1941), 595–598.

[6] *Sur une nouvelle démonstration d'un théorème d'Ahlfors.* Ibid. 212 (1941), 662–665.

[7] *Sur certaines propriétés nouvelles des fonctions algébroides.* Ibid. 212 (1941), 746–749.

[8] *Sur les fonctions méromorphes à caractéristique bornée.* Ibid. 213 (1941), 393–395.

[9] *Sur quelques progrès récents de la théorie des fonctions d'une variable complexe.* Rev. Sci. 79 (1941), 608–612.

[10] *Sur l'aire sphérique décrite par les valeurs d'une fonction méromorphe.* Bull. Sci. Math. (2) 65 (1941), 214–219.

[11] *Sur les domaines couverts par les valeurs d'une fonction méromorphe ou algébroide.* Ann. Sci. École Norm. Sup. (3) 58 (1941), 179–259.

[12] *Sur les cercles de remplissage des fonctions méromorphes.* C.R. Acad. Sci. Paris 214 (1942), 467–469.

[13] *Une propriété des surfaces de recouvrement.* Ibid. 215 (1942), 252–253.

[14] *Un critère de famille normale.* Ibid. 215 (1942), 294–296.

[15] *Sur quelques propriétés des cercles de remplissage des fonctions méromorphes.* Ann. Sci. École Norm. Sup. (3) 59 (1942), 187–209.

[16] *Remarques sur les fonctions méromorphes dans le voisinage d'un point singulier essentiel isolé.* Bull. Soc. Math. France 70 (1942), 40–45.

[17] *Sur les valeurs ramifiées des fonctions méromorphes.* Ibid. 72 (1944), 76–92.

[18] *Sur les fonctions méromorphes dans le cercle unité et couvrant une aire bornée.* C.R. Acad. Sci. Paris 219 (1944), 274–276.

[19] *Sur les fonctions méromorphes et univalentes dans le cercle unité.* Bull. Sci. Math. (2) 69 (1945), 21–36.

[20] *Le problème des coefficients pour certaines fonctions méromorphes dans le cercle unité.* Ann. Acad. Sci. Fenn. Ser. A I No. 250/9 (1958), 7 pp.

DUFRESNOY, J.; PISOT, C.

[1] *Étude de certaines fonctions méromorphes bornées sur le cercle unité. Application à un ensemble fermé d'entiers algébriques.* Ann. Sci. École Norm. Sup. (3) 72 (1955), 69–92.

DUGUÉ, D.

[1] *Le défaut au sens de M. Nevanlinna dépend de l'origine choisie.* C.R. Acad. Sci. Paris 225 (1947), 555–556.

[2] *Sur les valeurs exceptionnelles de Julia et un problème qu'elles soulèvent.* Ibid. 233 (1951), 841–842.

Dugué, D.
[3] *Vers un théorème de Picard global.* Ann. Sci. École Norm. Sup. (3) 69 (1952), 65–81.
[4] *Note sur l'article précédent.* Ibid. (3) 72 (1955), 163.

Dwivedi, S. H.
[1] *Proximate orders and distribution of a-points of meromorphic functions.* Compositio Math. 15 (1963), 192–202.

Edrei, A.
[1] *Meromorphic functions with three radially distributed values.* Trans. Amer. Math. Soc. 78 (1955), 276–293.
[2] *Meromorphic functions with values that are both deficient and asymptotic.* Studies in mathematical analysis and related topics. Stanford Univ. Press, Stanford, Calif., 1962, pp. 93–103.
[3] *The deficiencies of meromorphic functions of finite lower order.* Duke Math. J. 31 (1964), 1–21.

Edrei, A.; Fuchs, W. H. J.
[1] *On the growth of meromorphic functions with several deficient values.* Trans. Amer. Math. Soc. 93 (1959), 292–328.
[2] *Valeurs déficientes et valeurs asymptotiques des fonctions méromorphes.* Comment. Math. Helv. 33 (1959), 258–295.
[3] *The deficiencies of meromorphic functions of order less than one.* Duke Math. J. 27 (1960), 233–249.
[4] *Bounds for the number of deficient values of certain classes of meromorphic functions.* Proc. London Math. Soc. (3) 12 (1962), 315–344.
[5] *On meromorphic functions with regions free of poles and zeros.* Acta Math. 108 (1962), 113–145.
[6] *Entire and meromorphic functions*, Prentice-Hall Series in Modern Analysis (to appear).

Edrei, A.; Fuchs, W. H. J.; Hellerstein, S.
[1] *Radial distribution and deficiencies of the values of a meromorphic function.* Pacific J. Math. 11 (1961), 135–151.

Edrei, A.; Shah, S. M.
[1] *A conjecture of R. Nevanlinna concerning the genus of a meromorphic function.* Proc. Amer. Math. Soc. 11 (1960), 319–323.

Eggleston, H. G.
[1] *The range set of a function meromorphic in the unit circle.* Proc. London Math. Soc. (3) 5 (1955), 500–512.

Elfving, G.
[1] *Zur Flächenstruktur und Wertverteilung. Ein Beispiel.* Acta Acad. Abo. Math. Phys. 8, No. 10 (1935), 13 pp.

Emig, P.
[1] *Meromorphic functions and the capacity function on abstract Riemann surfaces.* Doctoral dissertation, Univ. of California, Los Angeles, 1962. 71 pp.

Evans, G. C.
[1] *Potentials and positively infinite singularities of harmonic functions.* Monatsh. Math. Phys. 43 (1936), 419–424.

Fan, W.-K.
[1] *Sur les fonctions méromorphes quasi-exceptionnelles.* Bull. Sci. Math. (2) 61 (1937), 369–379.
[2] *Sur quelques classes de fonctions méromorphes quasi-exceptionnelles.* Sci. Record (N.S.) 3 (1959), 1–5.

Ferrand, J.
[1] *Sur les fonctions holomorphes ou méromorphes dans une couronne.* C.R. Acad. Sci. Paris 214 (1942), 50–52.

FLORACK, H.

[1] *Reguläre und meromorphe Funktionen auf nicht geschlossenen Riemannschen Flächen*. Schr. Math. Inst. Univ. Münster, No. 1 (1948), 34 pp.

FROSTMAN, O.

[1] *Über den Kapazitätsbegriff und einen Satz von R. Nevanlinna*. Medd. Lunds Univ. Mat. Sem. 1 (1934), 14 pp.

[2] *Über die defekten Werte einer meromorphen Funktion*. 8. Scand. Congr. Math. Stockholm 1934, pp. 392–396.

[3] *Potential d'équilibre et capacité des ensembles avec quelques applications à la théorie des fonctions*. Medd. Lunds Univ. Mat. Sem. 3 (1935), 115 pp.

[4] *Über die defekten Werte einer meromorphen Funktion*. Ibid. 4 (1939), 5 pp.

FUCHS, W. H. J.

[1] *A theorem on the Nevanlinna deficiencies of meromorphic functions of finite order*. Ann. of Math. (2) 68 (1958), 203–209.

[2–7] Cf. EDREI, A.; FUCHS, W. H. J. [1–6].

[8] Cf. EDREI, A.; FUCHS, W. H. J.; HELLERSTEIN, S. [1].

FULLER, D. J. H.

[1] *Mappings of bounded characteristic into arbitrary Riemann surfaces*. Pacific J. Math. 14 (1964), 895–915.

GEHRING, F. W.

[1] *The asymptotic values for analytic functions with bounded characteristic*. Quart. J. Math. Oxford Ser. (2) 9 (1958), 282–289.

GELFOND, A.

[1] *Sur le théorème de M. Picard*. C.R. Acad. Sci. Paris 188 (1929), 1536–1537.

GERST, I.

[1] *Meromorphic functions with simultaneous multiplication and addition theorems*. Trans. Amer. Math. Soc. 61 (1947), 469–481.

GHERMANESCU, M.

[1] *Sur le théorème de Picard-Borel*. Ann. Sci. École Norm. Sup. (3) 52 (1935), 221–268.

[2] *Une inégalité pour les algébroides*. Bull. Math. Phys. École Polytech. Bucarest 10 (1938–39), 31–33.

GOL'DBERG, A. A.

[1] *On the inverse problem of the theory of the distribution of the values of meromorphic functions*. Ukrain. Mat. Ž. 6 (1954), 385–397. (Russian.)

[2] *On the influence of clustering of algebraic branch points of a Riemann surface on the order of growth of a meromorphic mapping function*. Dokl. Akad. Nauk SSSR 98 (1954), 709–711; Correction 101 (1955), 4. (Russian.)

[3] *On defects of meromorphic functions*. Ibid. 98 (1954), 893–895. (Russian.)

[4] *An estimate of the sum of the defects of a meromorphic function of order less than unity*. Ibid. 114 (1957), 245–248. (Russian.)

[5] *On an inequality connected with logarithmic convex functions*. Dopovidi Akad. Nauk Ukrain, RSR (1957), 227–230. (Ukrainian.)

[6] *On the set of defective values of meromorphic functions of finite order*. Ukrain. Mat. Ž. 11 (1959), 438–443. (Russian.)

[7] *Meromorphic functions with separated zeros and poles*. Izv. Vysš. Učebn. Zaved. Matematika No. 4 (17) (1960), 67–72. (Russian.)

[8] *Contemporary research on the Nevanlinna theory of distribution of values of meromorphic functions of finite order*. Issledovanija po sovremennym problemam teorii funkcií kompleksnogo peremennogo, pp. 406–417. Gosudarstv. Izdat. Fiz.-Mat. Lit., Moscow, 1960. (Russian.)

[9] *Distribution of values of meromorphic functions with separated zeros and poles*. Dokl. Akad. Nauk SSSR 137 (1961), 1030–1033. (Russian.) Translated as Soviet Math. Dokl. 2, 389–392.

[10] Cf. BELINSKIĬ, P. P.; GOL'DBERG, A. A. [1].

Gol'dberg, A. A.; Ostrovskiĭ, I. V.
[1] *New investigations on the growth and distribution of values of entire and meromorphic functions of genus zero.* Uspehi. Mat. Nauk 16 No. 4 (100) (1961), 51–62. (Russian.)

Goldstein, M.
[1] *L- and K-kernels on arbitrary Riemann surfaces.* Doctoral dissertation, Univ. of California, Los Angeles, 1962. 71 pp.

Habetha, K.
[1] *Über die Werteverteilung in Winkelräumen.* Math. Z. 77 (1961), 453–467.

af Hällström, G.
[1] *Über eindeutige analytische Funktionen mit unendlich vielen wesentlichen Singularitäten.* 9. Scand. Congr. Math. Helsinki 1938, pp. 277–284.
[2] *Über meromorphe Funktionen mit mehrfach zusammenhängenden Existenzgebieten.* Acta Acad. Abo. Math. Phys. 12 No. 8 (1939), 100 pp.
[3] *Eine quasikonforme Abbildung mit Anwendungen auf die Wertverteilungslehre.* Ibid. 18 No. 8 (1952), 16 pp.
[4] *Zur Berechnung der Bodenordnung oder Bodenhyperordnung eindeutiger Funktionen.* Ann. Acad. Sci. Fenn. Ser. A I No. 193 (1955), 16 pp.
[5] *Übertragung eines Satzkomplexes von Weierstrass und Dinghas auf beliebige Randmengen der Kapazität Null.* Ibid. Ser. A I No. 250/12 (1958), 9 pp.
[6] *Wertverteilungssätze pseudomeromorpher Funktionen.* Acta Acad. Abo. Math. Phys. 21 No. 9 (1958), 23 pp.
[7] *A new approach to the first fundamental theorem on value distribution.* Michigan Math. J. 9 (1962), 241–248.
[8] *On the main theorems of the distribution of values.* Nordisk Mat. Tidskr. 12 (1964), 53–67, 88. (Swedish. English summary.)

Halmos, P. R.
[1] *Measure theory.* D. Van Nostrand Company, Inc., Princeton, N.J., 1950. 304 pp.

Hayman, W. K.
[1] *The maximum modulus and valency of functions meromorphic in the unit circle.* Acta Math. 86 (1951), 89–191, 193–257.
[2] *On Nevanlinna's second theorem and extensions.* Rend. Circ. Mat. Palermo (2) 2 (1953), 346–392.
[3] *Picard values of meromorphic functions and their derivatives.* Ann. of Math. (2) 70 (1959), 9–42.
[4] *Meromorphic functions.* Clarendon Press, Oxford, 1964. 191 pp.
[5] *On the characteristic of functions meromorphic in the unit disk and of their integrals.* Acta Math. 112 (1964), 181–214.

Heins, M.
[1] *Riemann surfaces of infinite genus.* Ann. of Math. (2) 55 (1952), 296–317.
[2] *On the Lindelöf principle.* Ibid. (2) 61 (1955), 440–473.
[3] *Lindelöfian maps.* Ibid. (2) 62 (1955), 418–446.
[4] *Functions of bounded characteristic and Lindelöfian maps.* Internat. Congr. Math. Edinburgh 1958, pp. 376–388.
[5] *On the boundary behavior of a conformal map of the open unit disk into a Riemann surface.* J. Math. Mech. 9 (1960), 573–581.

Hellerstein, S.
[1] *On a class of meromorphic functions with deficient zeros and poles.* Pacific J. Math. 13 (1963), 115–124.
[2] Cf. Edrei, A.; Fuchs, W. H. J.; Hellerstein, S. [1].

Hellerstein, S.; Rubel, L. A.
[1] *Subfields that are algebraically closed in the field of all meromorphic functions.* J. Analyse Math. 12 (1964), 105–111.

HERVÉ, M.

[1] *À propos d'un mémoire récent de M. Noshiro: Nouvelles applications de sa méthode.* C.R. Acad. Sci. Paris 232 (1951), 2170–2172.

[2] *Sur les valeurs omises par une fonction méromorphe.* Ibid. 240 (1955), 718–720.

[3] *Contribution à l'étude d'une fonction méromorphe au voisinage d'un ensemble singulier de capacité nulle.* J. Math. Pures Appl. (9) 35 (1956), 161–173.

[4] *Valeurs exceptionnelles d'une fonction méromorphe au voisinage d'un ensemble singulier de capacité nulle.* Ann. Acad. Sci. Fenn. Ser. A I No. 250/14 (1958), 4 pp.

HERZ, J.

[1] *Über meromorphe transzendente Funktionen auf Riemannschen Flächen.* Doctoral dissertation, München. B. G. Teubner, Leipzig, 1933. 41 pp.

HIONG, K.-L.

[1] *Sur les fonctions méromorphes d'ordre infini.* C.R. Acad. Sci. Paris 196 (1933), 239–242.

[2] *Sur les fonctions méromorphes dans le cercle-unité.* Ibid. 196 (1933), 1764–1767.

[3] *Sur les fonctions entières et les fonctions méromorphes d'ordre infini.* J. Math. Pures Appl. (9) 14 (1935), 233–308.

[4] *Some properties of the meromorphic functions of infinite order.* Sci. Rep. Tsing Hua Univ. 3 (1935), 1–25.

[5] *Sur une extension du second théorème fondamental de R. Nevanlinna.* C.R. Acad. Sci. Paris 230 (1950), 1635–1636.

[6] *Sur les fonctions méromorphes et leurs dérivées.* Ibid. 231 (1950), 323–325.

[7] *Sur les fonctions holomorphes dans le cercle unité ne prenant une valeur que p fois et admettant une valeur exceptionnelle au sens de Picard-Borel ou au sens de R. Nevanlinna.* Ibid. 236 (1953), 1628–1630.

[8] *Un théorème général relatif à la croissance des fonctions holomorphes et privées de zéros dans le cercle unité et un nouveau critère de normalité pour une famille de fonctions holomorphes ou méromorphes.* Ibid. 236 (1953), 1322–1324.

[9] *Généralisations du théorème fondamental de Nevanlinna-Milloux.* Bull. Sci. Math. (2) 78 (1954), 181–198.

[10] *Un théorème d'unicité relatif à la théorie des fonctions méromorphes.* C.R. Acad. Sci. Paris 241 (1955), 1691–1693.

[11] *Un théorème fondamental sur les fonctions méromorphes et leurs primitives.* Ibid. 242 (1956), 53–55.

[12] *Sur la croissance des fonctions algébroides en rapport avec leurs dérivées.* Ibid. 242 (1956), 3032–3035.

[13] *Sur la limitation de $T(r, f)$ sans intervention des pôles.* Bull. Sci. Math. (2) 80 (1956), 175–190.

[14] *Sur les fonctions algébroides et leurs dérivées. Étude des défauts absolus et des défauts relatifs.* Ann. Sci. École Norm. Sup. (3) 73 (1956), 439–451.

[15] *Sur les fonctions méromorphes et les fonctions algébroides. Extensions d'un théorème de M. R. Nevanlinna.* Gauthier-Villars, Paris, 1957. 104 pp.

[16] *Sur les fonctions méromorphes en rapport avec leurs dérivées.* Sci. Sinica 7 (1958), 661–685.

[17] *Quelques théorèmes sur les fonctions méromorphes admettant un ensemble de valeurs déficientes.* Sci. Record (N.S.) 3 (1959), 61–64.

[18] *On the limitation of a meromorphic function admitting exceptional values B.* Bul. Inst. Politehn. Iași (N.S.) 5 (9) No. 3–4 (1959), 1–4. (Russian and Romanian summaries.)

[19] *Sur les fonctions méromorphes en rapport avec leurs primitives.* J. Math. Pures Appl. (9) 39 (1960), 1–31.

[20] *Sur la théorie des défauts relative aux fonctions méromorphes dans le cercle unité.* Sci. Sinica 9 (1960), 575–603.

[21] *Inégalités relatives à une fonction méromorphe et à l'une de ses primitives. Applications.* J. Math. Pures Appl. (9) 41 (1962), 1–34.

HIONG, K.-L.; HO, Y.

[1] *Sur les valeurs multiples des fonctions méromorphes et de leurs dérivées.* Sci. Sinica 10 (1961), 267–285.

[2] *On the multiple values of a meromorphic function and its derivative.* Chinese Math. 3 (1963), 156–168.

HONG, I.

[1] *On positively infinite singularities of a solution of the equation* $\Delta u + k^2 u = 0$. Kōdai Math. Sem. Rep. 8 (1956), 9–12.

HUCKEMANN, F.

[1] *Über den Einfluss von Randstellen Riemannscher Flächen auf die Wertverteilung.* Math. Z. 65 (1956), 240–282.

[2] *Über den Defekt von mittelbaren Randstellen auf beschränktartigen Riemannschen Flächen.* Ann. Acad. Sci. Fenn. Ser. A I No. 250/16 (1958), 12 pp.

INOUE, M.

[1] *Positively infinite singularities of solutions of linear elliptic partial differential equations.* J. Inst. Polytech. Osaka City Univ. Ser. A. 8 (1957), 43–50.

IVERSEN, F.

[1] *Recherches sur les fonctions inverses des fonctions méromorphes.* Doctoral dissertation, Univ. Helsinki, 1914. 67 pp.

JENKINS, J. A.

[1] *Sur quelques aspects globaux du théorème de Picard.* Ann. Sci. École Norm. Sup. (3) 72 (1955), 151–161.

JENSEN, J. L.

[1] *Sur un nouvel et important théorème de la théorie des fonctions.* Acta Math. 22 (1899), 359–364.

JØRGENSEN, V.

[1] *Über einen Zusatz zum Picard-Landauschen Satz.* Mat. Tidsskr. B (1939), 1–6. (Danish.)

JULIA, G.

[1] *Sur quelques propriétés nouvelles des fonctions entières ou méromorphes. I, II, III.* Ann. Sci. École Norm. Sup. (3) 36 (1919), 93–125; 37 (1920), 165–218; 38 (1921), 165–181.

[2] *Leçons sur les fonctions uniformes à point singulier essentiel isolé.* Gauthier-Villars, Paris, 1924. 151 pp.

KAKUTANI, S.

[1] *On the function m(r, a) in the theory of meromorphic functions.* Japan. J. Math. 13 (1937), 393–404.

[2] Cf. SHIMIZU, T.; YOSIDA, K.; KAKUTANI, S. [1].

KAMETANI, S.

[1] *The exceptional values of functions with the set of linear measure zero of essential singularities. I, II.* Proc. Imp. Acad. Tokyo 17 (1941), 117–120; 19 (1943), 438–443.

[2] *The exceptional values of functions with the set of capacity zero of essential singularities.* Ibid. 17 (1941), 429–433.

[3] *On Hausdorff's measures and generalized capacities with some of their applications to the theory of functions.* Japan. J. Math. 19 (1945), 217–257.

KELLEY, J. L.

[1] *General Topology.* D. Van Nostrand Company, Inc., Princeton, N.J., 1955. 298 pp.

KISHI, M.

[1] *Maximum principles in the potential theory.* Nagoya Math. J. 23 (1963), 165–187.

KLOTZ, T.; SARIO, L.
[1] *Existence of complete minimal surfaces of arbitrary connectivity and genus.* Proc. Nat. Acad. Sci. U.S.A. 54 (1965), 42–44.
[2] *Gaussian mapping of arbitrary minimal surfaces.* J. Analyse Math. 17 (1966), 209–218.

KNESER, H.; ULLRICH, E.
[1] *Funktionentheorie.* Naturforschung und Medizin in Deutschland 1939–1946, Band 1, pp. 189–242. Dieterich'sche Verlagsbuchhandlung, Wiesbaden, 1948.

KOMATU, Y.
[1] *Note on the theory of conformal representation by meromorphic functions. I.* Proc. Japan. Acad. 21 (1945), 269–277.

KOSEKI, K.
[1] *Über die Ausnahmewerte der meromorphen Funktionen.* Mem. Coll. Sci. Univ. Kyoto Ser. A Math. 27 (1952), 7–40.

KRASNER, M.
[1] *Essai d'une théorie des fonctions analytiques dans les corps valués complets: théorèmes de Nevanlinna: transformations holomorphes.* C.R. Acad. Sci. Paris 222 (1946), 363–365.

KRAWTCHOUK, M.
[1] *Sur les pôles des fonctions méromorphes.* C.R. Acad. Sci. Paris 185 (1927), 178–179.

KUNUGUI, K.
[1] *Sur un problème de M. A. Beurling.* Proc. Imp. Acad. Tokyo 16 (1940), 361–366.
[2] *Une généralisation des théorèmes de MM. Picard-Nevanlinna sur les fonctions méromorphes.* Ibid. 17 (1941), 283–288.
[3] *Sur l'allure d'une fonction analytique uniforme au voisinage d'un point frontière de son domaine de définition.* Japan. J. Math. 18 (1942), 1–39.
[4] *Sur la théorie de la distribution des valeurs.* Proc. Imp. Acad. Tokyo 18 (1942), 269–275.
[5] *Sur la théorie des fonctions méromorphes et uniformes.* Japan. J. Math. 18 (1943), 583–614.

KÜNZI H. P.
[1] *Über ein Teichmüllersches Wertverteilungsproblem.* Arch. Math. 4 (1953), 210–215.
[2] *Neue Beiträge zur geometrischen Wertverteilungslehre.* Comment. Math. Helv. 29 (1955), 223–257.

KÜNZI, H. P.; WITTICH, H.
[1] *Sur la répartition des points où certaines fonctions méromorphes prennent une valeur a.* C.R. Acad. Sci. Paris 245 (1957), 1991–1994.
[2] *The distribution of the a-points of certain meromorphic functions.* Michigan Math. J. 6 (1959), 105–121.

KURAMOCHI, Z.
[1] *On covering surfaces.* Osaka Math. J. 5 (1953), 155–201.
[2] *Evans' theorem on abstract Riemann surfaces with null boundaries. I, II.* Proc. Japan Acad. 32 (1956), 1–6; 7–9.
[3] *Mass distributions on the ideal boundaries of abstract Riemann surfaces. I.* Osaka Math. J. 8 (1956), 119–137.

KURODA, T.
[1] *On the uniform meromorphic functions with the set of capacity zero of essential singularities.* Tôhoku Math. J. (2) 3 (1951), 257–269.
[2] *On analytic functions on some Riemann surfaces.* Nagoya Math. J. 10 (1956), 27–50.

KUSUNOKI, Y.
[1] *On the property of Riemann surfaces and the defect.* Mem. Coll. Sci. Univ. Kyoto Ser. A Math. 26 (1950), 63–73.

LANDAU, E.

[1] *Über eine Verallgemeinerung des Picardschen Satzes.* S.-B. Preuss. Akad. Wiss. (1904), 1118–1133.

[2] *Über den Picardschen Satz.* Zürich. Naturf. Ges. 51 (1906), 252–318.

LANDAU, H. J.

[1] Cf. OSSERMAN, R.; LANDAU, H. J. [1].

LANDAU, R.

[1] Cf. BOHR, H.; LANDAU, R. [1].

LEE, K.

[1] *Über die Verallgemeinerung einiger Ergebnisse der Wertverteilungstheorie der meromorphen Funktionen.* Acta Math. Sinica 3 (1953), 87–100.

LEE, K. P.

[1] *On meromorphic functions of infinite order.* Japan. J. Math. 12 (1935), 1–16; 37–42.

[2] *On the unified theory of meromorphic functions.* Proc. Phys.-Math. Soc. Japan (3) 18 (1936), 182–187.

[3] *On the Borel's directions of meromorphic functions of infinite order.* Japan. J. Math. 13 (1936), 39–48.

[4] *On the directions of Borel of meromorphic functions of positive finite order.* Publ. Dep. Math. Sun-Yat Sen Univ. 1 (1937), 439–452.

[5] *On the unified theory of meromorphic functions.* J. Fac. Sci. Univ. Tokyo I 3 (1937), 253–286.

[6] *Sur les directions de Borel des fonctions méromorphes d'ordre fini supérieur à 1/2.* C.R. Acad. Sci. Paris 206 (1938), 811–812.

[7] *Sur les directions de Borel des fonctions méromorphes d'ordre infini.* Ibid. 206 (1938), 1548–1550.

[8] *Sur les valeurs multiples et les directions de Borel des fonctions méromorphes.* Ibid. 206 (1938), 1784–1786.

[9] *On the directions of Borel of meromorphic functions of finite order* $> 1/2$. Compositio Math. 6 (1938), 285–295.

LEHTO, O.

[1] *Sur la théorie des fonctions méromorphes à caractéristique bornée.* C.R. Acad. Sci. Paris 236 (1953), 1943–1945.

[2] *On meromorphic functions whose values lie in a given domain.* Ann. Acad. Sci. Fenn. Ser. A I No. 160 (1953), 15 pp.

[3] *A majorant principle in the theory of functions.* Math. Scand. 1 (1953), 5–17.

[4] *On an extension of the concept of deficiency in the theory of meromorphic functions.* Ibid. 1 (1953), 207–212.

[5] *On meromorphic functions of bounded characteristic.* 12. Scand. Congr. Math. Lund 1953, pp. 183–187.

[6] *Value distribution and boundary behaviour of a function of bounded characteristic and the Riemann surface of its inverse function.* Ann. Acad. Sci. Fenn. Ser. A I No. 177 (1954), 46 pp.

[7] *On the distribution of values of meromorphic functions of bounded characteristic.* Acta Math. 91 (1954), 87–112.

[8] *Boundary theorems for analytic functions.* Ann. Acad. Sci. Fenn. Ser. A I No. 196 (1955), 8 pp.

[9] *Distribution of values and singularities of analytic functions.* Ibid. Ser. A I No. 249/3 (1957), 16 pp.

[10] *A generalization of Picard's theorem.* Ark. Mat. 3 (1958), 495–500.

[11] *The spherical derivative of meromorphic functions in the neighborhood of an isolated singularity.* Comment. Math. Helv. 33 (1959), 196–205.

LEHTO, O.; VIRTANEN, K. I.

[1] *Boundary behaviour and normal meromorphic functions.* Acta Math. 97 (1957), 47–65.

LEHTO, O; VIRTANEN, K. I.

[2] *On the behaviour of meromorphic functions in the neighborhood of an isolated singularity.* Ann. Acad. Sci. Fenn. Ser. A I No. 240 (1957), 9 pp.

LELONG, P.

[1] *Sur le principe de Lindelöf et les valeurs asymptotiques d'une fonction méromorphe d'ordre fini.* C.R. Acad. Sci. Paris 204 (1937), 652–654.

[2] *Sur la capacité de certains ensembles de valeurs exceptionnelles.* Ibid. 214 (1942), 992–994.

LEVINE, H. I.

[1] *A theorem on holomorphic mappings into complex projective space.* Ann. of Math. 71 (1960), 529–535.

LINDELÖF, E.

[1] *Sur le théorème de M. Picard dans la théorie des fonctions monogènes.* Scand. Congr. Math. Stockholm 1909, pp. 112–136.

LOHWATER, A. J.

[1] *The exceptional values of meromorphic functions.* Colloq. Math. 7 (1959), 89–93.

LOOMIS, L. H.

[1] *An introduction to abstract harmonic analysis.* D. Van Nostrand Company, Inc., Princeton, N.J., 1953. 190 pp.

MACINTYRE, A. J.

[1] *A theorem concerning meromorphic functions of finite order.* Proc. London Math. Soc. (2) 39 (1935), 282–294.

MACINTYRE, A. J.; SHAH, S. M.

[1] *On an extension of a theorem of Bernstein to meromorphic functions.* J. Math. Anal. Appl. 3 (1961), 351–354.

MACLANE, G. R.

[1] *Meromorphic functions with small characteristic and no asymptotic values.* Michigan Math. J. 8 (1961), 177–185.

MÄDER, O.

[1] *Über das asymptotische Verhalten meromorpher Funktionen bei speziel gegebener Null- und Polstellenverteilung.* Doctoral dissertation, Univ. of Freiburg, Switzerland, 1942. 27 pp.

MALLIAVIN, P.

[1] *Sur la croissance radiale d'une fonction méromorphe.* Illinois J. Math. 1 (1957), 259–296.

MANJANATHAIAH, K.

[1] Cf. SINGH, S. K.; MANJANATHAIAH, K. [1].

MARTY, F.

[1] *Sur la répartition des valeurs d'une fonction méromorphe.* C.R. Acad. Sci. Paris 190 (1930), 466–468.

[2] *Recherches sur la répartition des valeurs d'une fonction méromorphe.* Ann. Fac. Sci. Univ. Toulouse (3) 23 (1931), 183–261.

MATHUR, Y. B. L.

[1] *On exceptional values of meromorphic functions. I, II, III, IV.* Proc. Nat. Acad. Sci. India Sect. A 21 (1952), 213–216; 217–219; 220–223; 224.

MATSUMOTO, K.

[1] *Exceptional values of meromorphic functions in a neighborhood of the set of singularities.* J. Sci. Hiroshima Univ. Ser. A 24 (1960), 143–153.

[2] *On exceptional values of meromorphic functions with the set of singularities of capacity zero.* Nagoya Math. J. 18 (1961), 171–191.

[3] *Remark on Lehto's paper: "A generalisation of Picard's theorem."* Proc. Japan Acad. 38 (1962), 636–640.

MATSUMOTO, K.

[4] *Some notes on exceptional values of meromorphic functions.* Nagoya Math. J. 22 (1963), 189–201.

[5] *Existence of perfect Picard set.* Ibid. 27 (1965), 213–222.

MAZURKIEWICZ, S.

[1] *Sur une classe de fonctions méromorphes.* C.R. Soc. Sci. Lett. Varsovie Cl. III 29 (1936), 7–9.

MEIMANN, N.

[1] *Sur les pôles des fonctions méromorphes.* Commun. Soc. Math. Kharkoff (4) 14 (1937), 97–104. (Ukranian. French summary.)

MIKHAIL, M. N.

[1] *On the a-values of the random meromorphic function.* Nederl. Akad. Wetensch. Proc. Ser. A 59=Indag. Math. 18 (1956), 170–180.

[2] *The behaviour of the random meromorphic function at its zeros. I, II.* Ibid. Ser. A 60=Indag. Math. 19 (1957), 88–95; 96–103.

[3] *The behaviour of the random meromorphic function at its poles.* Ibid. Ser. A 60=Indag. Math. 19 (1957), 590–597.

MIKOLÁS, M.

[1] *Bemerkungen über den Wertvorrat durch Potenzreihen definierter, insbesondere meromorpher Funktionen.* Ann. Univ. Sci. Budapest. Eötvös. Sect. Math. 2 (1959), 123–132.

MILLOUX, H.

[1] *Le théorème de M. Picard, suites de fonctions holomorphes, fonctions méromorphes et fonctions entières.* J. Math. Pures Appl. (9) 3 (1924), 345–401.

[2] *Les fonctions méromorphes à valeur asymptotique et le théorème de M. Picard.* C.R. Acad. Sci. Paris 180 (1925), 809–812.

[3] *Sur le théorème de Picard.* Bull. Soc. Math. France 53 (1926), 181–207.

[4] *Sur quelques propriétés des racines des fonctions méromorphes.* C.R. Acad. Sci. Paris 186 (1928), 933–934.

[5] *Sur quelques propriétés des fonctions méromorphes et holomorphes.* Ibid. 189 (1929), 896–898.

[6] *Les cercles de remplissage des fonctions méromorphes ou entières et le théorème de Picard-Borel.* Acta Math. 52 (1929), 189–255.

[7] *Remarques sur la théorie des fonctions méromorphes.* Proc. Phys.-Math. Soc. Japan (3) 12 (1930), 9–21.

[8] *Sur les valeurs asymptotiques des fonctions entières d'ordre infini.* Compositio Math. 1 (1934), 305–313.

[9] *Sur quelques points de la théorie des fonctions méromorphes dans un cercle.* Internat. Congr. Math. Oslo 1936, 2, pp. 68–69.

[10] *Étude des fonctions méromorphes dans un cercle.* C.R. Acad. Sci. Paris 202 (1936), 1480–1482.

[11] *Sur les fonctions méromorphes dans un cercle.* Ibid. 204 (1937), 1394–1395.

[12] *Sur la théorie des fonctions méromorphes dans le cercle unité.* Ann. Sci. École Norm. Sup. (3) 54 (1937), 151–229.

[13] *Fonctions méromorphes.—Contributions à l'étude des ensembles de points où ces fonctions sont proches d'une valeur donnée.* J. Math. Pures Appl. (9) 16 (1937), 179–198.

[14] *Fonctions méromorphes dans un cercle.* Ibid. (9) 17 (1938), 257–274.

[15] *Une inégalité nouvelle dans la théorie des fonctions méromorphes.* C.R. Acad. Sci. Paris 208 (1939), 31–32.

[16] *Sur la théorie des défauts.* Ibid. 210 (1940), 38–39.

[17] *Les fonctions méromorphes et leurs dérivées. Extensions d'un théorème de M. R. Nevanlinna. Applications.* Hermann et Cie., Paris, 1940. 53 pp.

[18] *Sur une nouvelle extension d'une inégalité de M. R. Nevanlinna.* J. Math. Pures Appl. (9) 19 (1940), 197–210.

MILLOUX, H.

[19] *Sur une inégalité de M. R. Nevanlinna.* Revista Ci. Lima 47 (1945), 507–544.

[20] *Les dérivées des fonctions méromorphes et la théorie des défauts.* Ann. Sci. École Norm. Sup. (3) 63 (1947), 289–316.

[21] *Sur quelques propriétés des fonctions méromorphes et de leurs dérivées.* C.R. Acad. Sci. Paris 234 (1952), 39–41.

[22] *Sur une propriété des fonctions méromorphes et de leurs dérivées.* J. Math. Pures Appl. (9) 31 (1952), 1–18.

MILNOR, J.

[1] *Morse theory.* Princeton Univ. Press, Princeton, N.J., 1963. 153 pp.

MIZUMOTO, H.

[1] *A note on an abelian covering surface. I.* Kōdai Math. Sem. Rep. 15 (1963), 29–51.

MONTEL, P.

[1] *Leçons sur les fonctions entières ou méromorphes.* Gauthier-Villars, Paris, 1932. 116 pp.

[2] *Sur les valeurs algèbriques d'une fonction entière ou méromorphe.* J. Math. Pures Appl. (9) 20 (1941), 305–324.

[3] *Sur les propriétés tangentielles des fonctions analytiques.* Rev. Math. Pures Appl. 3 (1958), 5–8.

[4] *Éléments exceptionnels des fonctions analytiques.* Ann. Sci. École Norm. Sup. 76 (1959), 271–281.

MORI, A.

[1] *Valiron's theorem on Picard's curves.* Kōdai Math. Sem. Rep. (1950), 101–103.

[2] *A note on unramified abelian covering surfaces of a closed Riemann surface.* J. Math. Soc. Japan 6 (1954), 162–176.

MYRBERG, L.

[1] *Über meromorphe Funktionen und Kovarianten auf Riemannschen Flächen.* Ann. Acad. Sci. Fenn. Ser. A I No. 244 (1957), 18 pp.

[2] *Eine Bemerkung zum Picardschen Satz.* Ibid. Ser. A I No. 255 (1958), 4 pp.

[3] *Über einige Extremalgrössen in der Theorie der meromorphen Funktionen.* Ibid. Ser. A I No. 284 (1960), 9 pp.

[4] *Über meromorphe Funktionen auf endlich vielblättrigen Riemannschen Flächen. I, II, III.* Ibid. Ser. A I No. 286 (1960), 17 pp.; No. 301 (1961), 9 pp.; No. 311 (1962), 7 pp.

[5] *Über meromorphe Funktionen auf nullberandeten Riemannschen Flächen.* Ibid. Ser. A I No. 312 (1962), 11 pp.

MYRBERG, P. J.

[1] *Über beschränkte Funktionen in mehrfach zusammenhängenden Bereichen.* Ann. Acad. Sci. Fenn. Ser. A I 33 No. 8 (1930), 15 pp.

[2] *L'existence de la fonction de Green pour un domaine plan donné.* C.R. Acad. Sci. Paris 190 (1930), 1372–1374.

[3] *Über die Existenz der Greenschen Funktionen auf einer gegebenen Riemannschen Fläche.* Acta Math. 61 (1933), 39–79.

[4] *Über die analytische Fortsetzung von beschränkten Funktionen.* Ann. Acad. Sci. Fenn. Ser. A I No. 58 (1949), 7 pp.

NAKAI, M.

[1] *On Evans potential.* Proc. Japan Acad. 38 (1962), 624–629.

[2] *Evans' harmonic functions on Riemann surfaces.* Ibid. 39 (1963), 74–78.

[3] *On Evans' solution of the equation $\Delta u = Pu$ on Riemann surfaces.* Kōdai Math. Sem. Rep. 15 (1963), 79–93.

[4] *On the fundamental existence theorem of Kishi.* Nagoya Math. J. 23 (1963), 189–198.

[5] *Green Potential of Evans type on Royden's compactification of a Riemann surface.* Ibid. 24 (1964), 205–239.

[6] *Potentials of Sario's kernel.* J. Analyse Math. 17 (1966), 225–240.

NEVANLINNA, F.
[1] *Über die Anwendung einer Klasse uniformisierender Transzendenten zur Untersuchung der Wertverteilung analytischer Funktionen.* Acta Math. 50 (1927), 159–188.
[2] *Über die logarithmische Ableitung einer meromorphen Funktion.* Ann. Acad. Sci. Fenn. 32 No. 12 (1929), 11 pp.
[3] *Über eine Klasse meromorpher Funktionen.* 7. Scand. Congr. Math. Oslo 1929, pp. 81–83.

NEVANLINNA, F.; NEVANLINNA, R.
[1] *Über die Eigenschaften analytischer Funktionen in der Umgebung einer singulären Stelle oder Linie.* Acta Soc. Sci. Fenn. 50 No. 5 (1922), 46 pp.

NEVANLINNA, R.
[1] *Beweis des Picard-Landauschen Satzes.* Nachr. Acad. Wiss. Göttingen Math.-Phys. Kl.II. (1924), 151–154.
[2] *Untersuchungen über den Picardschen Satz.* Acta Soc. Sci. Fenn. 50 No. 6 (1924), 42 pp.
[3] *Über den Picard-Borelschen Satz in der Theorie der ganzen Funktionen.* Ann. Acad. Sci. Fenn. 23 No. 5 (1924), 37 pp.
[4] *Über eine Klasse meromorpher Funktionen.* Math. Ann. 92 (1924), 145–154.
[5] *Zur Theorie der meromorphen Funktionen.* Acta Math. 46 (1925), 1–99.
[6] *Un théorème d'unicité relatif aux fonctions uniformes dans le voisinage d'un point singulier essentiel.* C.R. Acad. Sci. Paris 181 (1925), 92–94.
[7] *Quelques propriétés des fonctions méromorphes dans un angle donné.* Ibid. 181 (1925), 352–354.
[8] *Über die Eigenschaften meromorpher Funktionen in einem Winkelraum.* Acta Soc. Sci. Fenn. 50 No. 12 (1925), 45 pp.
[9] *Neuere Untersuchungen über den Picardschen Satz.* 6. Scand. Congr. Math. Copenhagen 1925, pp. 77–95.
[10] *Über die Werteverteilung einer analytischen Funktion in der Umgebung einer isolierten wesentlich singulären Stelle.* Ibid. pp. 97–107.
[11] *Einige Eindeutigkeitssätze in der Theorie der meromorphen Funktionen.* Acta Math. 48 (1926), 367–391.
[12] *Sur les valeurs exceptionnelles des fonctions méromorphes dans un cercle.* Bull. Soc. Math. France 55 (1927), 92–101.
[13] *Compléments aux théorèmes d'unicité dans la théorie des fonctions méromorphes.* C.R. Acad. Sci. Paris 186 (1928), 289–291.
[14] *Sur les théorèmes d'unicité dans la théorie des fonctions uniformes.* Internat. Congr. Math. Bologna 1928, 3, pp. 223–228.
[15] *Le théorème de Picard-Borel et la théorie des fonctions méromorphes.* Gauthier-Villars, Paris, 1929. 174 pp.
[16] *Über gewisse neuere Ergebnisse in der Theorie der Wertverteilung.* 7. Scand. Congr. Math. Oslo 1929, pp. 68–80.
[17] *Über die Herstellung transzendenter Funktionen als Grenzwerte rationaler Funktionen.* Acta Math. 55 (1930), 259–276.
[18] *Über die Randwerte von analytischen Funktionen.* Comment. Math. Helv. 2 (1930), 236–252.
[19] *Sur une classe de fonctions transcendantes.* C.R. Acad. Sci. Paris 191 (1930), 914–916.
[20] *Über die Werteverteilung der eindeutigen analytischen Funktionen.* Abh. Math. Sem. Univ. Hamburg 8 (1931), 351–400.
[21] *Über Riemannsche Flächen mit endlich vielen Windungspunkten.* Acta Math. 58 (1932), 295–373.
[22] *Eindeutige analytische Funktionen.* Springer-Verlag, Berlin-Göttingen-Heidelberg. 1te Aufl. 1936. 353 pp. 2te Aufl. 1953. 379 pp.
[23] *Über die Existenz von beschränkten Potentialfunktionen auf Flächen von unendlichem Geschlecht.* Math. Z. 52 (1950), 599–604.

NINOMIYA, N.

[1] *Étude sur la théorie du potentiel pris par rapport au noyau symétrique.* J. Inst. Polytech. Osaka City Univ. Ser. A 8 (1957), 147–179.

NOSHIRO, K.

[1] *Contributions to the theory of meromorphic functions in the unit-circle.* J. Fac. Sci. Hokkaido Univ. Ser. I 7 (1939), 149–159.

[2] *On the singularities of analytic functions.* Japan. J. Math. 17 (1940), 37–96.

[3] *On the singularities of analytic functions with a general domain of existence.* Proc. Japan Acad. 22 No. 8 (1946), 233–237.

[4] *Contributions to the theory of the singularities of analytic functions.* Japan. J. Math. 19 No. 4 (1948), 299–327.

[5] *Open Riemann surface with null boundary.* Nagoya Math. J. 3 (1951), 73–79.

[6] *The modern theory of functions.* Iwanami shoten, Tokyo, 1954. 428 pp.

[7] *On the theory of cluster sets of analytic functions.* Amer. Math. Soc. Transl. (2) 8 (1958), 1–12.

[8] *Cluster sets.* Springer-Verlag, Berlin-Göttingen-Heidelberg, 1960. 135 pp.

NOSHIRO, K.; SARIO, L.

[1] *Integrated forms derived from nonintegrated forms of value distribution theorems under analytic and quasi-conformal mappings.* Festschrift zur Gedächtnisfeier für Karl Weierstrass 1815–1965. Wissenschaftliche Abhandlungen der Arbeitsgemeinschaft für Forschung des Landes Nordrhein-Westfalen 33 (1966), 319–324. Westdeutscher Verlag, Köln und Opladen.

OGAWA, S.; SAKAGUCHI, K.

[1] *Some classes of meromorphic functions with assigned zeros and poles.* J. Math. Soc. Japan 8 (1956), 40–53.

OĞUZTÖRELI, M. N.

[1] *Sur une généralisation de la formule de Jensen et quelques applications.* Rev. Fac. Sci. Univ. Istanbul Ser. A 15 (1950), 289–332.

[2] *Extension de la théorie de Nevanlinna aux domaines multiplement connexes.* Ibid. Ser. A 18 (1953), 384–419.

[3] *Représentations intégrales de la fonction caractéristique, de la fonction de nombre et de la forme sphérique normale generalisée et extension d'un théorème de Borel.* Ibid. Ser. A 19 (1954), 79–85.

[4] *Sur les propriétés relatives à la distribution des valeurs qui correspondent aux solutions des équations différentielles linéaires à coéfficients elliptiques.* Ibid. Ser. A 22 (1957), 91–96.

[5] *Sur les fonctions à type borné.* Ibid. Ser. A 22 (1957), 141–149.

[6] *Sur une classe des fonctions méromorphes définies par une équation différentielle.* Atatürk Univ. Yayinlari Ser.-Mat. No. 1 (1959), 90 pp. (Turkish and French.)

OHTSUKA, M.

[1] *Dirichlet problems on Riemann surfaces and conformal mappings.* Nagoya Math. J. 3 (1951), 91–137.

[2] *On exceptional values of a meromorphic function.* Ibid. 9 (1955), 119–121.

[3] *Reading of the paper "On covering surfaces" by Z. Kuramochi.* Chapter I (mimeographed).

OIKAWA, K.

[1] *Sario's lemma on harmonic functions.* Proc. Amer. Math. Soc. 11 (1960), 425–428.

OKADA, Y.

[1] *Some theorems on meromorphic functions.* Proc. Amer. Math. Soc. 1 (1950), 246–249.

OSGOOD, W. F.

[1] *Meromorphe Funktionen in einem beliebig erweiterten Raum.* Leopoldina 4 (1929), 126–128.

OSSERMAN, R.
[1] *Proof of a conjecture of Nirenberg*. Comm. Pure Appl. Math. 12 (1959), 229–232.
[2] *Global properties of minimal surfaces in E^3 and E^n*. Ann. of Math. 80 (1964), 340–364.
[3] *Minimal surfaces in the large*. Comment. Math. Helv. 35 (1961), 65–76.
[4] *Global properties of classical minimal surfaces*. Duke Math. J. 32 (1965), 565–573.

OSSERMAN, R.; LANDAU, H. J.
[1] *On analytic mappings of Riemann surfaces*. J. Analyse Math. 7 (1959/60), 249–279.

OSTROVSKIĬ, I. V.
[1] *On meromorphic functions taking certain values at points lying near a finite system of rays*. Dokl. Akad. Nauk SSSR 120 (1958), 970–972. (Russian.)
[2] *An estimate for the defect of a meromorphic function for which two values are distributed inside a certain angle*. Izv. Vysš. Učebn. Zaved. Matematika No. 2 (15) (1960), 138–148. (Russian.)
[3] *On the relation of the growth of a meromorphic function to the distribution of its values according to their arguments*. Dokl. Akad. Nauk SSSR 132 (1960), 48–51. (Russian.) Translated as Soviet Math. Dokl. 1, 485–488.
[4] *The connection between the growth of a meromorphic function and the distribution of the arguments of its values*. Izv. Akad. Nauk SSSR Ser. Mat. 25 (1961), 277–328. (Russian.)
[5] *Deficiencies of meromorphic functions of order less than one*. Dokl. Akad. Nauk SSSR 150 (1963), 32–35. (Russian.)
[6] *A problem from the theory of distribution of values*. Ibid. 151 (1963), 34–37. (Russian.)
[7] Cf. GOL'DBERG, A. A.; OSTROVSKIĬ, I. V. [1].

OSTROWSKI, A.
[1] *Über Folgen analytischer Funktionen und einige Verschärfungen des Picardschen Satzes*. Math. Z. 24 (1925), 215–258.

OU, V. T.
[1] *Valeurs déficientes d'une fonction algébroide*. C.R. Acad. Sci. Paris 232 (1951), 2073–2075.

OZAWA, M.
[1] *Picard's theorem on some Riemann surfaces*. Kōdai Math. Sem. Rep. 15 (1963), 245–256.
[2] *On the growth of analytic functions*. Ibid. 16 (1964), 98–100.
[3] *Remarks on unramified abelian covering surfaces of a closed Riemann surface*. Ibid. 16 (1964), 101–104.
[4] *Rigidity of projection map and the growth of analytic functions*. Ibid. 16 (1964), 40–43.
[5] *On the existence of analytic mappings*. Ibid. 17 (1965), 191–197.
[6] *On complex analytic mappings*. Ibid. 17 (1965), 93–102.
[7] *On ultrahyperelliptic surfaces*. Ibid. 17 (1965), 103–108.
[8] *On complex analytic mappings between two ultrahyperelliptic surfaces*. Ibid. 17 (1965), 158–165.

PARREAU, M.
[1] *Variation du défaut d'Ahlfors avec l'origine du plan des z*. C.R. Acad. Sci. Paris 227 (1948), 1198–1199.
[2] *Sur les moyennes des fonctions harmoniques et analytiques et la classification des surfaces de Riemann*. Ann. Inst. Fourier Grenoble 3 (1952), 103–197.
[3] *Fonction caractéristique d'une application conforme*. Ann. Fac. Sci. Univ. Toulouse (4) 19 (1955), 175–190.
[4] *Fonction caractéristique d'une application conforme. Relation avec la notion d'application de type Bl*. C.R. Acad. Sci. Paris 241 (1955), 1545–1546.

PETRENKO, V. P.

[1] *The growth of meromorphic functions along a ray.* Dokl. Akad. Nauk SSSR 155 (1964), 281–284. (Russian.)

[2] *Some estimates for the logarithmic derivative of a meromorphic function.* Izv. Akad. Nauk Armjan. SSR Ser. Fiz.-Mat. Nauk 17 (1964), No. 1, 23–27. (Russian. Armenian summary.)

[3] *On the deficiencies of a meromorphic function.* Dokl. Akad. Nauk SSSR 158 (1964), 1030–1033. (Russian.)

PETROVITCH, M.

[1] *Propositions sur les fonctions méromorphes.* Publ. Math. Univ. Belgrade 5 (1936), 163–168.

PFLUGER, A.

[1] *Zur Defektrelation ganzer Funktionen endlicher Ordnung.* Comment. Math. Helv. 19 (1946), 91–104.

[2] *Sur l'existence de fonctions non constantes, analytiques, uniformes et bornées sur une surface de Riemann ouverte.* C.R. Acad. Sci. Paris 230 (1950), 166–168.

PICARD, E.

[1] *Sur une propriété des fonctions entières.* C.R. Acad. Sci. Paris 88 (1879), 1024–1027.

[2] *Mémoire sur les fonctions entières.* Ann. Sci. École Norm. Sup. (2) 9 (1880), 147–166.

[3] *Démonstration d'un théorème général sur les functions uniformes liées par une relation algébrique.* Acta Math. 11 (1887), 1–12.

PIRANIAN, G.

[1] Cf. COLLINGWOOD, E. F.; PIRANIAN, G. [1].

PISOT, C.

[1] Cf. DUFRESNOY, J.; PISOT, C. [1].

PÖSCHL, K.

[1] *Über die Wertverteilung der erzeugenden Funktionen Riemannscher Flächen mit endlich vielen periodischen Enden.* Math. Ann. 123 (1951), 79–95.

PRINGSHEIM, A.

[1] *Elementare Theorie der ganzen transzendenten Funktionen von endlicher Ordnung.* Math. Ann. 58 (1904), 257–342.

PU, P.-M.

[1] *On the unified theory of meromorphic functions in the unit circle.* Wu-Han Univ. J. Sci. 8 No. 1 (1942), 3.1–3.14.

RADOITCHITCH, M.

[1] *Sur les fonctions inverses des fonctions méromorphes.* C.R. Acad. Sci. Paris 189 (1929), 1240–1242.

RAJAGOPAL, C. T.

[1] *On periodic meromorphic functions.* J. Indian Math. Soc. (N.S.) 9 (1945), 69–76.

RAO, K. V. R.

[1] *Lindelöfian maps and positive harmonic functions.* Doctoral dissertation, Univ. of California, Los Angeles, 1962. 48 pp.

[2] *Lindelöfian meromorphic functions.* Proc. Amer. Math. Soc. 15 (1964), 109–113.

[3] *Remarks on the classification of Riemann surfaces.* Ibid. 15 (1964), 632–634.

RAUCH, A.

[1] *Généralisation de théorèmes de M. Valiron sur les fonctions méromorphes d'ordre positif.* C.R. Acad. Sci. Paris 192 (1931), 1189–1191.

[2] *Extensions de théorèmes relatifs aux directions de Borel des fonctions méromorphes.* J. Math. Pures Appl. (9) 12 (1933), 109–171.

RÉMOUNDOS, G.
[1] *Sur l'extension du théorème de M. Picard aux fonctions multiformes.* C.R. Acad. Sci. Paris 181 (1925), 459–461.
[2] *Sur un cas d'élimination et l'extension aux fonctions algébroides du théorème de M. Picard.* Ann. Mat. Pura Appl. (4) 2 (1925), 107–110.
[3] *Extensions aux fonctions algébroides multiformes du théorème de M. Picard et de ses généralisations.* Gauthier-Villars, Paris, 1927. 66 pp.
[4] *Sur la nouvelle généralisation du théorème de M. Picard.* Verh. Akad. Athens 3 (1928), 60–62.

RIIBER, A. E.
[1] *Über meromorphe Funktionen mit einem Existenzgebiete, dessen Rand eine Cantor'sche Punktmenge von der Kapazität Null ist.* Math. Scand. 3 (1955), 229–242.

ROBINSON, R. M.
[1] *A generalization of Picard's and related theorems.* Duke Math. J. 5 (1939), 118–132.

RODIN, B.
[1] *Reproducing formulas on Riemann surfaces.* Doctoral dissertation, Univ. of California, Los Angeles, 1961. 71 pp.
[2] *The sharpness of Sario's generalized Picard theorem.* Proc. Amer. Math. Soc. 15 (1964), 373–374.

RODIN, B.; SARIO, L.
[1] *Existence of mappings into noncompact Riemann surfaces.* J. Analyse Math. 17 (1966), 219–224.

ROYDEN, H. L.
[1] *Rings of analytic and meromorphic functions.* Trans. Amer. Math. Soc. 83 (1956), 269–276.

RUBEL, L. A.
[1] Cf. HELLERSTEIN, S.; RUBEL, L. A. [1].

RUDIN, W.
[1] *Positive infinities of potentials.* Proc. Amer. Math. Soc. 2 (1951), 967–970.

SAGAWA, A.
[1] *Über die Ausnahmegebiete.* Math. Japonicae 2 (1952), 146–148.

SĂGINYAN, A. L.
[1] *On some inequalities and their applications in the theory of functions.* Dokl. Akad. Nauk SSSR 129 (1959), 284–287. (Russian.)

SAKAGUCHI, K.
[1] Cf. OGAWA, S.; SAKAGUCHI, K. [1].

SARIO, L.
[1] *A linear operator method on arbitrary Riemann surfaces.* Trans. Amer. Math. Soc. 72 (1952), 281–295.
[2] *Modular criteria on Riemann surfaces.* Duke Math. J. 20 (1953), 279–286.
[3] *Capacity of the boundary and of a boundary component.* Ann. of Math. 59 (1954), 135–144.
[4] *Picard's great theorem on Riemann surfaces.* Amer. Math. Monthly 69 (1962), 598–608.
[5] *Meromorphic functions and conformal metrics on Riemann surfaces.* Pacific J. Math. 12 (1962), 1079–1097.
[6] *Analytic mappings between arbitrary Riemann surfaces.* (Research Announcement.) Bull. Amer. Math. Soc. 68 (1962), 633–637.
[7] *Islands and peninsulas on arbitrary Riemann surfaces.* Trans. Amer. Math. Soc. 106 (1963), 521–533.

SARIO, L.

[8] *On locally meromorphic functions with single-valued moduli.* Pacific J. Math. 13 (1963), 709–724.

[9] *Value distribution under analytic mappings of arbitrary Riemann surfaces.* Acta Math. 109 (1963), 1–10.

[10] *General value distribution theory.* Nagoya Math. J. 23 (1963), 213–229.

[11] *Second main theorem without exceptional intervals on arbitrary Riemann surfaces.* Michigan Math. J. 10 (1963), 207–219.

[12] *Complex analytic mappings.* Bull. Amer. Math. Soc. 69 (1963), 439–445.

[13] *An integral equation and a general existence theorem for harmonic functions.* Comment. Math. Helv. 38 (1964), 284–292.

[14] *A theorem on mappings into Riemann surfaces of infinite genus.* Trans. Amer. Math. Soc. 117 (1965), 276–284.

[15] Cf. AHLFORS, L.; SARIO, L. [1].

[16–17] Cf. KLOTZ, T.; SARIO, L. [1–2].

[18] Cf. NOSHIRO, K.; SARIO, L. [1].

[19] Cf. RODIN, B.; SARIO, L. [1]

SAXER, W.

[1] *Sur les valeurs exceptionnelles des dérivées successives des fonctions méromorphes.* C.R. Acad. Sci. Paris 182 (1926), 831–833.

[2] *Sur les fonctions méromorphes quasi-exceptionnelles.* Ibid. 184 (1927), 264–266.

[3] *Über quasi-normale Funktionsscharen und eine Verschärfung des Picardschen Satzes.* Math. Ann. 99 (1928), 707–737.

SCHMIDT, E.

[1] *Über die isoperimetrische Aufgabe im n-dimensionalen Raum konstanter negativer Krümmung.* Math. Z. 46 (1940), 204–230.

[2] *Die isoperimetrischen Ungleichungen auf der gewöhnlichen Kugel und für Rotationskörper im n-dimensionalen sphärischen Raum.* Ibid. 46 (1940), 743–794.

SCHOTTKY, F.

[1] *Über den Picardschen Satz und die Borelschen Ungleichungen.* S.-B. Preuss. Akad. Wiss., Physik.-Math. Kl. (1904), 1244–1263.

[2] *Über zwei Beweise des allgemeinen Picardschen Satzes.* Ibid. (1907), 823–840.

SCHUBART, H.

[1] *Zur Wertverteilung der Painlevéschen Transzendenten.* Arch. Math. 7 (1956), 284–290.

[2] *Zur Wertverteilung der elliptischen Funktionen und der Painlevéschen Transzendenten als Lösungen algebraischer Differentialgleichungen.* Univ. e Politec. Torino Rend. Sem. Mat. 17 (1957/58), 161–186.

SCHWARTZ, M.-H.

[1] *Sur une propriété de la fonction m(r, A) de M. Nevanlinna dans les fonctions méromorphes.* C.R. Acad. Sci. Paris 210 (1940), 525–526.

[2] *Exemple d'une fonction méromorphe ayant des valeurs déficientes non asymptotiques.* Ibid. 212 (1941), 382–384.

[3] *Sur les indices de ramification de M. Nevanlinna.* Ibid. 228 (1949), 45–46.

[4] *Formules apparentées à celles de Nevanlinna-Ahlfors pour certaines applications d'une variété à n dimensions dans une autre.* Bull. Soc. Math. France 82 (1954), 317–360.

SĚDA, V.

[1] *A note to a paper of Clunie.* Acta Fac. Nat. Univ. Comenian. 4 (1959), 255–260.

SEIDEL, W.

[1] *On the distribution of values of bounded analytic functions.* Trans. Amer. Math. Soc. 36 (1934), 201–226.

[2–3] Cf. BAGEMIHL, F.; SEIDEL, W. [1–2].

SELBERG, H. L.
[1] *Sur le théorème de M. Picard.* C.R. Acad. Sci. Paris 187 (1928), 100–102.
[2] *Über einige Eigenschaften bei der Wertverteilung der meromorphen Funktionen endlicher Ordnung.* Avh. Norske Vid.-Akad. Oslo No. 7 (1928), 17 pp.
[3] *Über eine Eigenschaft der logarithmischen Ableitung einer meromorphen oder algebroiden Funktion endlicher Ordnung.* Ibid. I No. 14 (1929), 11 pp.
[4] *Bemerkungen zur Theorie der algebroiden Funktionen.* 7. Scand. Congr. Math. Oslo 1929, pp. 89–91.
[5] *Über die Wertverteilung der algebroiden Funktionen.* Math. Z. 31 (1930), 709–728.
[6] *Beiträge zur Theorie der algebroiden Funktionen.* Avh. Norske Vid.-Akad. Oslo No. 9 (1931), 22 pp.
[7] *Ein Satz aus der Theorie der algebroiden Funktionen.* Ibid. No. 8 (1932), 16 pp.
[8] *Eine Werteverteilungseigenschaft der algebroiden Funktionen.* Ibid. No. 14 (1932), 12 pp.
[9] *Algebroide Funktionen und Umkehrfunktionen Abelscher Integrale.* Ibid. No. 8 (1934), 72 pp.
[10] *Ein Existenzsatz der Potentialtheorie und seine Anwendung.* Ibid. No. 6 (1935), 10 pp.
[11] *Über die ebenen Punktmengen von der Kapazität Null.* Ibid. No. 10 (1937), 10 pp.
[12] *Über einen Darstellungssatz aus der Theorie der meromorphen Funktionen.* 9. Scand. Congr. Math. Helsinki 1938, pp. 62–66.
[13] *Eine Ungleichung der Potentialtheorie und ihre Anwendung in der Theorie der meromorphen Funktionen.* Comment. Math. Helv. 18 (1946), 309–326.
[14] *Über einen Satz von Picard.* Norske Vid. Selsk. Forh. (Trondheim) 36 (1963), 145–149.

SERGHIESCO, S.
[1] *Sur le nombre des zéros et des pôles distincts d'une fonction méromorphe dans un contour fermé.* C.R. Acad. Sci. Paris 225 (1947), 485–487.

SHAH, S. M.
[1] *A note on meromorphic functions.* Math. Student 12 (1945), 67–70.
[2] *Some theorems on meromorphic functions.* Proc. Amer. Math. Soc. 2 (1951), 694–698.
[3] *Exceptional values of entire and meromorphic functions.* Duke Math. J. 19 (1952), 585–593.
[4] *Exceptional values of entire and meromorphic functions. II.* J. Indian Math. Soc. (N.S.) 20 (1956), 315–327.
[5] *Meromorphic functions of finite order.* Proc. Amer. Math. Soc. 10 (1959), 810–821.
[6] *On exceptional values and the genus of a meromorphic function.* Math. Z. 75 (1960/61), 385–391.
[7] *On the order of the difference of two meromorphic functions.* Proc. Amer. Math. Soc. 12 (1961), 234–242.
[8] Cf. EDREI, A.; SHAH, S. M. [1].
[9] Cf. MACINTYRE, A. J.; SHAH, S. M. [1].

SHAH, S. M.; SINGH, S. K.
[1] *Borel's theorem on a-points and exceptional values of entire and meromorphic functions.* Math. Z. 59 (1953), 88–93.
[2] *On the derivative of a meromorphic function with maximum defect.* Ibid. 65 (1956), 171–174.
[3] *Meromorphic functions with maximum defect sum.* Tôhoku Math. J. (2) 11 (1959), 447–452.

SHANKAR, H.
[1] *On the characteristic function of a meromorphic function. I.* Tôhoku Math. J. (2) 9 (1957), 243–246.

SHANKAR, H.

[2] *A note on entire and meromorphic functions.* Publ. Math. Debrecen 5 (1958), 213–216.

[3] *Note on a theorem of Shah.* Rend. Circ. Mat. Palermo (2) 8 (1959), 225–227.

SHIMIZU, T.

[1] *On the theory of meromorphic functions.* Proc. Imp. Acad. Tokyo 5 (1929), 105–107.

[2] *On the theory of meromorphic functions.* Japan. J. Math. 6 (1929), 119–171.

[3] *Some theorems on meromorphic functions.* Proc. Phys.-Math. Soc. Japan (3) 11 (1929), 28–35.

[4] *Remarks on a proof of Picard's general theorem and allied theorems.* Japan. J. Math. 6 (1930), 315–318.

[5] *On the function-group for a meromorphic function. I, II.* Proc. Phys.-Math. Soc. Japan (3) 13 (1931), 297–301; (3) 14 (1932), 36–40.

[6] *On equi-modular functions and their applications.* Ibid. (3) 13 (1931), 79–92.

[7] *On the fundamental domain and the groups for meromorphic functions. I, II.* Japan. J. Math. 8 (1931), 175–236; 237–307.

[8] *On the linear functions of automorphism for meromorphic functions.* Tôhoku Math. J. 38 (1933), 219–224.

[9] *On the existence of meromorphic functions which are automorphic with respect to some function-groups.* Proc. Phys.-Math. Soc. Japan (3) 15 (1933), 433–448.

SHIMIZU, T.; YOSIDA, K.; KAKUTANI, S.

[1] *On meromorphic functions. I.* Proc. Phys.-Math. Soc. Japan (3) 17 (1935), 1–10.

SINGH, S. K.

[1] *A note on entire and meromorphic functions.* Proc. Amer. Math. Soc. 9 (1958), 6–10.

[2] *On exceptional values of entire and meromorphic functions.* Tôhoku Math. J. (2) 13 (1961), 373–380.

[3–5]. Cf. SHAH, S. M.; SINGH, S. K. [1–3].

SINGH, S. K.; MANJANATHAIAH, K.

[1] *On exceptional values of meromorphic functions.* Publ. Math. Debrecen 11 (1964), 18–22.

SOULA.

[1] *Sur les zéros et les pôles d'une fonction méromorphe dans un secteur.* C.R. Acad. Sci. Paris 198 (1934), 890–891.

SOULA, J.

[1] *Sur les fonctions méromorphes au voisinage d'un point singulier à l'intérieur.* Mathematica Cluj 10 (1935), 81–91.

SRIVASTAV, R. P.

[1] *On zeros, poles and mean values of meromorphic functions.* Compositio Math. 13 (1958), 219–228.

STARKE, E. P.

[1] *Certain uniform functions of rational functions.* Trans. Amer. Math. Soc. 29 (1927), 276–286.

STEIN, K.

[1] Cf. BEHNKE, H.; STEIN, K. [1].

STOLL, W.

[1] *Die beiden Hauptsätze der Wertverteilungstheorie bei Funktionen mehrerer komplexer Veränderlichen. I, II.* Acta Math. 90 (1953), 1–115; 92 (1954), 55–169.

STORVICK, D. A.

[1] *On meromorphic functions of bounded characteristic.* Proc. Amer. Math. Soc. 8 (1957), 32–38.

SUNYER BALAGUER, F.

[1] *Number of Borel directions and exceptional values of a meromorphic function of finite order.* Mem. Real Acad. Ci. Art. Barcelona 30 (1952), 451–459.

SURYANARAYANAN, K. S.

[1] *Composite meromorphic functions.* J. Indian Math. Soc. (2) 1 (1935), 241–246.

TAMURA, J.

[1] *Meromorphic functions on open Riemann surfaces.* Sci. Papers Coll. Gen. Ed. Univ. Tokyo 9 (1959), 175–186.

TEICHMÜLLER, O.

[1] *Eine Umkehrung des zweiten Hauptsatzes der Wertverteilungslehre.* Deutsche Math. 2 (1937), 96–107.

[2] *Vermutungen und Sätze über die Wertverteilung gebrochener Funktionen endlicher Ordnung.* Ibid. 4 (1939), 163–190.

[3] *Einfache Beispiele zur Wertverteilungslehre.* Ibid. 7 (1944), 360–368.

THIEM, L.-V.

[1] *Le degré de ramification d'une surface de Riemann et la croissance de la caractéristique de la fonction uniformisante.* C.R. Acad. Sci. Paris 228 (1949), 1192–1195.

[2] *Über das Umkehrproblem der Wertverteilungslehre.* Comment. Math. Helv. 23 (1949), 26–49.

[3] *Sur un problème d'inversion dans la théorie des fonctions méromorphes.* Ann. Sci. École Norm. Sup. (3) 67 (1950), 51–98.

TIETZ, H.

[1] *Zur Klassifizierung meromorpher Funktionen auf Riemannschen Flächen.* Math. Ann. 142 (1960/61), 441–449.

TÔKI, Y.

[1] *On the behaviour of a meromorphic function in the neighborhood of a transcendental singularity.* Proc. Imp. Acad. Tokyo 17 (1941), 296–300.

[2] *Proof of Ahlfors principal covering theorem.* Rev. Math. Pures Appl. 2 (1957), 277–280.

TROHIMČUK, YU. YU.

[1] *Sur la généralisation de théorème de Picard.* Ukrain. Mat. Ž. 10 No. 1 (1958), 70–77. (Russian. French summary.)

TSUJI, M.

[1] *On a generalization of Picard's theorem.* Proc. Imp. Acad. Tokyo 2 (1926), 364–365.

[2] *On the theorems of Valiron and Milloux.* Japan. J. Math. 15 (1939), 255–267.

[3] *On the behaviour of an inverse function of a meromorphic function at its transcendental singular point. I, II, III.* Proc. Imp. Acad. Tokyo 17 (1941), 414–417; 474–475; 18 (1942), 132–139.

[4] *On the behaviour of a meromorphic function in the neighborhood of a closed set of capacity zero.* Ibid. 18 (1942), 213–219.

[5] *Nevanlinna's fundamental theorems and Ahlfors' theorem on the number of asymptotic values.* Japan. J. Math. 18 (1943), 675–708.

[6] *On the Riemann surface of an inverse function of a meromorphic function in the neighborhood of a closed set of capacity zero.* Proc. Imp. Acad. Tokyo 19 (1943), 257–258.

[7] *Theory of meromorphic functions in a neighborhood of a closed set of capacity zero.* Japan. J. Math. 19 (1944), 139–154.

[8] *On Borel's directions of meromorphic functions of finite order. I, II, III.* Tôhoku Math. J. (2) 2 (1950), 97–112; Kōdai Math. Sem. Rep. (1950), 96–100; 104–108.

[9] *On a regular function which is of constant absolute value on the boundary of an infinite domain.* Tôhoku Math. J. (2) 3 (1951), 24–38.

TSUJI, M.

[10] *On meromorphic functions with essential singularities of logarithmic capacity zero.* Ibid. (2) 3 (1951), 1–6.

[11] *On the order of the derivative of a meromorphic function.* Ibid. (2) 3 (1951), 282–284.

[12] *On the remainder term of Nevanlinna's second fundamental theorem.* J. Math. Soc. Japan 4 (1952), 31–36.

[13] *On Ahlfors' theorems on covering surfaces.* J. Fac. Sci. Univ. Tokyo Sect. I 6 (1953), 319–328.

[14] *On a direct transcendental singularity of an inverse function of a meromorphic function.* J. Math. Soc. Japan 5 (1953), 75–80.

[15] *Theory of meromorphic functions on an open Riemann surface with null boundary.* Nagoya Math. J. 6 (1953), 137–150.

[16] *A remark on my former paper "Theory of Fuchsian groups."* J. Math. Soc. Japan 7 (1955), 202–207.

[17] *Borel's direction of a meromorphic function in a unit circle.* Ibid. 7 (1955), 290–311.

[18] *On the capacity of a set in the space of regular functions and its applications.* Comment. Math. Univ. St. Paul. 7 (1959), 1–12.

[19] *Potential theory in modern function theory.* Maruzen Co., Tokyo, 1959. 590 pp.

TUMURA, Y.

[1] *Sur quelques propriétés d'une classe simple des fonctions méromorphes.* Proc. Phys.-Math. Soc. Japan (3) 18 (1936), 173–181.

[2] *Sur les théorèmes de M. Valiron et les singularités transcendantes indirectement critiques.* Proc. Imp. Acad. Tokyo 17 (1941), 65–69.

[3] *Sur le premier théorème dans la théorie des fonctions méromorphes.* Ibid. 18 (1942), 164–169.

[4] *Quelques applications de la théorie de M. Ahlfors.* Japan. J. Math. 18 (1942), 303–322.

[5] *Recherches sur la distribution des valeurs des fonctions analytiques.* Ibid. 18 (1943), 797–876.

UGAERI, T.

[1] *On the general potential and capacity.* Japan. J. Math. 20 (1950), 37–43.

ULLRICH, E.

[1] *Über die Ableitung einer meromorphen Funktion.* S.-B. Deutsch. Akad. Wiss. Berlin Kl. Math. Phys. Tech. (1929), 592–608.

[2] *Über die Ausnahmewerte von algebroiden Funktionen.* Akad. Wiss. Wien Math.-Naturwiss. Kl. 68 (1931), 27–31.

[3] *Über eine Anwendung des Verzerrungssatzes auf meromorphe Funktionen.* J. Reine Angew. Math. 166 (1931), 220–234.

[4] *Über den Einfluss der Verzweightheit einer Algebroide auf ihre Wertverteilung.* Ibid. 167 (1932), 198–220.

[5] *Eine Abbildungsaufgabe zur Theorie der Wertverteilung.* Internat. Congr. Math. Zürich 1932, 2, pp. 45–46.

[6] *Zum Umkehrproblem der Wertverteilungslehre.* Nachr. Akad. Wiss. Göttingen Math.-Phys. Kl. FG. I 1 (1936), 135–150.

[7] *Über das Umkehrproblem der Wertverteilungslehre.* Internat. Congr. Math. Oslo 1936, 2, pp. 69–72.

[8] *Flächenbau und Wertverteilung.* 9. Scand. Congr. Math. Helsinki 1938, pp. 179–200.

[9] Cf. KNESER, H.; ULLRICH, E. [1].

UTZ, W. R.

[1] *On the decomposition of meromorphic functions.* Revista Ci., Lima 50 (1948), 167–170.

VALIRON, G.

[1] *Sur les fonctions entières d'ordre nul et d'ordre fini et en particulier les fonctions à correspondence régulière.* Ann. Fac. Sci. Univ. Toulouse (3) 5 (1914), 117–257.

[2] *Sur les zéros des fonctions entières d'ordre infini.* C.R. Acad. Sci. Paris 172 (1921), 741–744.

[3] *Remarque sur un théorème de M. Julia.* Bull. Sci. Math. (2) 49 (1925), 68–73.

[4] *Sur la distribution des valeurs des fonctions méromorphes.* Acta Math. 47 (1925), 117–142.

[5] *Supplément à la note "Remarque sur un théorème de M. Julia."* Bull. Sci. Math. (2) 49 (1925), 270–275.

[6] *Sur les valeurs asymptotiques de quelques fonctions méromorphes.* Rend. Circ. Mat. Palermo 49 (1925), 415–421.

[7] *Sur les fonctions méromorphes qui sont exceptionnelles relativement au théorème de M. Julia.* C.R. Acad. Sci. Paris 180 (1925), 1895–1896.

[8] *Fonctions entières et fonctions méromorphes d'une variable.* Gauthier-Villars, Paris, 1925. 58 pp.

[9] *Sur les fonctions méromorphes sans valeurs asymptotiques.* C.R. Acad. Sci. Paris 182 (1926), 1266–1268. Correction 1432.

[10] *Sur une propriété des fonctions méromorphes d'ordre positif.* Bull. Sci. Math. (2) 50 (1926), 168–174.

[11] *Sur les fonctions méromorphes qui admettent des valeurs quasi-exceptionnelles.* Assoc. Française Lyons 1926, pp. 82–85.

[12] *Compléments au théorème de Picard-Julia.* Bull. Sci. Math. (2) 51 (1927), 167–183.

[13] *Le théorème de M. Picard et le complément de M. Julia.* J. Math. Pures Appl. (9) 7 (1928), 113–126.

[14] *Un théorème général sur les fonctions méromorphes d'ordre positif.* C.R. Acad. Sci. Paris 186 (1928), 26–28.

[15] *Sur quelques propriétés des fonctions méromorphes.* Ibid. 186 (1928), 935–936.

[16] *Sur les cercles de remplissage des fonctions méromorphes.* Ibid. 186 (1928), 1189–1191.

[17] *Sur les valeurs d'une fonction méromorphe dans le voisinage d'une singularité.* Ibid. 187 (1928), 803–805.

[18] *Recherches sur le théorème de M. Borel dans la théorie des fonctions méromorphes.* Acta Math. 52 (1928), 67–92.

[19] *Sur les fonctions algébroides méromorphes du second degré.* C.R. Acad. Sci. Paris 189 (1929), 623–625.

[20] *Sur les fonctions algébroides méromorphes.* Ibid. 189 (1929), 729–731.

[21] *Sur quelques propriétés des fonctions algébroides.* Ibid. 189 (1929), 824–826.

[22] *Familles normales et quasi-normales de fonctions méromorphes.* Gauthier-Villars, Paris, 1929. 55 pp.

[23] *Sur la dérivée d'une fonction méromorphe et sur certaines équations fonctionnelles.* C.R. Acad. Sci. Paris 190 (1930), 1223–1225.

[24] *Sur une propriété générale des fonctions méromorphes.* Ibid. 192 (1931), 269–271.

[25] *Remarques sur le théorème de M. Borel dans la théorie des fonctions méromorphes* Ibid. 192 (1931), 476–478.

[26] *Sur les directions de Borel des fonctions méromorphes d'ordre fini.* J. Math. Pures Appl. (9) 10 (1931), 457–480.

[27] *Sur la dérivée des fonctions algébroides.* Bull. Soc. Math. France 59 (1931), 17–39.

[28] *Points de Picard et points de Borel des fonctions méromorphes dans un cercle.* Bull. Sci. Math. (2) 56 (1932), 10–32.

[29] *Le théorème de Borel-Julia dans la théorie des fonctions méromorphes.* Internat. Congr. Math. Zürich 1932, 1, pp. 270–279.

[30] *Sur quelques conséquences de théorèmes de M. Ahlfors.* C.R. Acad. Sci. Paris 194 (1932), 1790–1792.

[31] *Remarques sur les valeurs exceptionnelles des fonctions méromorphes.* Rend. Circ. Mat. Palermo 57 (1933), 71–86.

VALIRON, G.

[32] *Sur le nombre des singularités transcendantes des fonctions inverses d'une classe d'algébroides*. C.R. Acad. Sci. Paris 200 (1935), 713–715.

[33] *Sur les directions de Borel des fonctions méromorphes d'ordre nul*. Bull. Sci. Math. (2) 59 (1935), 298–320.

[34] *Sur les variations du module des fonctions entières ou méromorphes*. C.R. Acad. Sci. Paris 204 (1937), 33–35.

[35] *Sur les valeurs exceptionnelles des fonctions méromorphes et de leurs dérivées*. Hermann et Cie., Paris, 1937. 53 pp.

[36] *Sur les directions de Borel des fonctions méromorphes d'ordre infini*. C.R. Acad. Sci. Paris 206 (1938), 575–577.

[37] *Sur les directions de Borel des fonctions algébroides méromorphes d'ordre infini*. Ibid. 206 (1938), 735–737.

[38] *Directions de Borel des fonctions méromorphes*. Gauthier-Villars, Paris, 1938. 70 pp.

[39] *Sur le domaine couvert par les valeurs d'une fonction algébroide finie*. Bull. Sci. Math. (2) 64 (1940), 199–206.

[40] *Valeurs exceptionnelles et valeurs déficientes des fonctions méromorphes*. C.R. Acad. Sci. Paris 225 (1947), 556–558.

[41] *Sur l'interpolation par les fonctions méromorphes*. Proc. Math. Phys. Soc. Egypt 4 No. 1 (1949), 23–26.

[42] *Sur les valeurs déficientes des fonctions méromorphes d'ordre nul*. C.R. Acad. Sci. Paris 230 (1950), 40–42.

[43] *Sur les valeurs déficientes des fonctions algébroides méromorphes d'ordre nul*. J. Analyse Math. 1 (1951), 28–42.

[44] *Sur une classe de fonctions algébroides d'ordre nul*. Rend. Circ. Mat. Palermo (2) 1 (1952), 63–70.

[45] *Fonctions entières d'ordre fini et fonctions méromorphes. V*. Enseignement Math. (2) 5 (1959), 1–28.

[46] *Fonctions entières d'ordre fini et fonctions méromorphes*. Monographies de "L'Enseignement Mathématique," No. 8. Institut de Mathématiques, Univ. Genève, 1960. 150 pp.

VAROPOULOS, T.

[1] *Sur les valeurs exceptionnelles des fonctions dérivées et le théorème de M. Saxer*. Acta Math. 51 (1927), 23–29.

[2] *Sur les fonctions algébroides quotient de deux algébroides bornées*. Bull. Sci. Math. (2) 52 (1928), 225–231.

VIRTANEN, K. I.

[1] *Eine Bermerkung über die Anwendung hyperbolischer Massbestimmungen in der Wertverteilungslehre der meromorphen Funktionen*. Math. Scand. 1 (1953), 153–158.

[2–3] Cf. LEHTO, O.; VIRTANEN, K. I. [1–2].

VOSS, K.

[1] *Über vollständige Minimalflächen*. Enseignement Math. 10 (1964), 316–317.

WAGNER, H.

[1] *Über eine Klasse Riemannscher Flächen mit endlich vielen nur logarithmischen Windungspunkten*. J. Reine Angew. Math. 175 (1936), 6–49.

WALSH, J. L.

[1] *An expansion of meromorphic functions*. Proc. Nat. Acad. Sci. U.S.A. 18 (1932), 165–171.

[2] *An interpolation series expansion for a meromorphic function*. Trans. Amer. Math. Soc. 74 (1953), 1–9.

WEIGAND, L.

[1] *Über die Randwerte meromorpher Funktionen einer Veränderlichen*. Comment. Math. Helv. 22 (1949), 125–149.

WEILL, G.

[1] *Reproducing kernels and orthogonal kernels for analytic differentials on Riemann surfaces*. Pacific J. Math. 12 (1962), 729–767.

[2] *Capacity differentials on open Riemann surfaces*. Ibid. 12 (1962), 769–776.

[3] *Some extremal properties of linear combinations of kernels on Riemann surfaces*. Ibid. 12 (1962), 1459–1465.

WEYL, H.

[1] *Meromorphic functions and analytic curves*. Princeton Univ. Press, Princeton, N.J., 1943. 269 pp.

WEYL, H.; WEYL, J.

[1] *Meromorphic curves*. Ann. of Math. (2) 39 (1938), 516–538.

[2] *On the theory of analytic curves*. Proc. Nat. Acad. Sci. U.S.A. 28 (1942), 417–421.

WEYL, J.

[1] *Analytic curves*. Ann. of Math. (2) 42 (1941), 371–408.

[2] *Exponential curves*. Duke Math. J. 10 (1943), 123–143.

WHITTAKER, J.-M.

[1] *On the fluctuation of integral and meromorphic functions. I*. Proc. London Math. Soc. (2) 37 (1934), 383–400.

[2] *The order of the derivative of a meromorphic function*. J. London Math. Soc. 11 (1936), 82–87.

WILLE, R. J.

[1] *On the integration of Ahlfors' inequality concerning covering surfaces*. Nederl. Akad. Wetensch. Proc. Ser. A 60 = Indag. Math. 19 (1957), 108–111.

WIMAN, A.

[1] *Sur le cas d'exception dans la théorie des fonctions entières*. Ark. Mat. 1 (1903), 327–345.

[2] *Sur une extension d'un théorème de M. Hadamard*. Ibid. 2 No. 14 (1903), 5 pp.

WISHARD, A.

[1] *Functions of bounded type*. Duke Math. J. 9 (1942), 663–676.

WITTICH, H.

[1] *Über eine Klasse meromorpher Funktionen*. Arch. Math. 1 (1948), 160–166.

[2] *Über den Einfluss algebraischer Windungspunkte auf die Wachstumsordnung*. Math. Ann. 122 (1950), 37–46.

[3] *Bemerkung zur Wertverteilung von Exponentialsummen*. Arch. Math. 4 (1953), 202–209.

[4] *Einige Eigenschaften der Lösungen von $w' = a(z) + b(z)w + c(z)w^2$*. Ibid. 5 (1954), 226–232.

[5] *Neuere Untersuchungen über eindeutige analytische Funktionen*. Springer-Verlag, Berlin-Göttingen-Heidelberg, 1955. 163 pp.

[6] *Über eine Klasse Riemannscher Flächen*. Comment. Math. Helv. 30 (1956), 116–123.

[7] *Defektfreie Lösungen linearer Differentialgleichungen*. Arch. Math. 7 (1957), 459–464.

[8] *Über die Ableitung einer meromorphen Funktion mit maximaler Defektsumme*. Math. Z. 69 (1958), 237–238.

[9] *Defekte Werte eindeutiger analytischer Funktionen*. Arch. Math. 9 (1958), 65–74.

[10] *Neuere Untersuchungen über eindeutige analytische Funktionen*. Gosudarstv. Izdat. Fiz.-Mat. Lit., Moscow, 1960. 319 pp. (Russian. Translated from the German and supplemented by A. A. Gol'dberg.)

[11–12] Cf. KÜNZI, H. P.; WITTICH, H. [1–2].

YAMAMURA, Y.

[1] *On a boundary theorem on open Riemann surfaces*. Proc. Japan Acad. 39 (1963), 17–20.

YOSIDA, K.

[1] *On a class of meromorphic functions.* Proc. Phys.-Math. Soc. Japan (3) 16 (1934), 227–235.

[2] *Corrigendum: "On a class of meromorphic functions."* Ibid. (3) 16 (1934), 413.

[3] Cf. SHIMIZU, T.; YOSIDA, K.; KAKUTANI, S. [1].

YÛJÔBÔ, Z.

[1] *An application of Ahlfors's theory of covering surfaces.* J. Math. Soc. Japan 4 (1952), 59–61.

SUBJECT INDEX

J stands for Introduction, A for Appendix.
Italicized section numbers refer to definitions.

AUTHOR INDEX

J stands for the Introduction to the book or to a chapter or section. A refers to Appendix.